Tree Management in Farmer Strategies

Tree Management in Farmer Strategies:

Responses to Agricultural Intensification

Edited by

J. E. Michael Arnold
and
Peter A. Dewees

Oxford Forestry Institute
University of Oxford

Oxford   New York   Tokyo   Melbourne
OXFORD UNIVERSITY PRESS
1995

Oxford University Press, Walton Street, Oxford OX2 6DP
Oxford New York Toronto
Delhi Bombay Calcutta Madras Karachi
Kuala Lumpur Singapore Hong Kong Tokyo
Nairobi Dar es Salaam Cape Town
Melbourne Auckland Madrid
and associated companies in
Berlin Ibadan

Oxford is a trade mark of Oxford University Press

Published in the United States
by Oxford University Press Inc., New York

A catalogue record for this book is available from the British Library

Library of Congress Cataloging in Publication Data
Tree management in farmer strategies : responses to agricultural intensification /
edited by J. E. Michael Arnold and Peter A. Dewees.
1. Tree crops. 2. Woodlots—Management. 3. Tree crops—South Asia.
4. Woodlots—Management—South Asia. 5. Tree crops—Africa, Eastern.
6. Woodlots—Management—Africa, Eastern. I. Arnold, J. E. M.
II. Dewees, Peter A. III. Title: Agricultural intensification.
SB170.T723   1994   338.1'7499'00954—dc20   94-25814
ISBN 0 19 858414 8

Typeset by The Electronic Book Factory Ltd, Fife, Scotland
Printed in Great Britain by
Biddles Ltd, Guildford and King's Lynn

Preface

In most parts of the developing world, rural people maintain trees in their vicinity in order to provide one or more locally important goods and services, including foods, fuel, fodder, building materials, saleable commodities and protection of the soil and crops. As natural forests nearby recede or degrade, farmers in many situations have historically tried to protect, plant and manage trees on their land in order to provide selected of these outputs. There are few farming systems today which do not incorporate managed trees in some fashion or another.

However, until recently this in-aggregate large, but widely dispersed, tree resource present within agricultural landscapes has attracted very little attention. Agricultural services were concerned only with those tree species that had been domesticated and adopted as agricultural crops, and forest services focused just on trees within areas defined as forests. Falling between the two, most tree stocks maintained by farmers remained effectively ignored, and even unnoticed.

As developmental thinking and practice came increasingly to focus on the rural sector, and on meeting basic needs and mobilizing rural resources, recognition of the importance of tree resources and forest outputs grew. In particular, the sharply increased concern with energy supplies following the 1973 jump in fossil energy prices drew attention to the extent to which people in the developing world depend on wood as their main fuel for cooking and other household needs. Early analyses emphasized the huge numbers of people affected, the apparent 'gap' between demand for fuelwood and sustainable supplies, the damaging environmental impacts of excessive removal of tree cover to meet burgeoning demands for fuelwood, and the adverse impacts on the poor of consequent shortages of cooked food and building materials, and of declining agricultural productivity as tree cover was removed and crop residues and dung were diverted from agricultural to fuel use.

There was widespread agreement at the time that one of the principal means of averting damaging shortages, and tree removal, was through the planting of additional trees in areas where the fuelwood was needed, and with the involvement of fuelwood users. As recognition grew of other benefits that might accrue from the presence of trees within farming systems, the impetus for programmes and projects to encourage tree planting by farmers increased rapidly. However, the subsequent design and implementation of interventions intended to support such actions have been hampered by lack of information about why farmers do or do not grow trees. By comparison with the crop and livestock components of agriculture very little is known about existing tree management practices, about farmers' perceptions of the value of trees and of different tree outputs in meeting their needs and production objectives, and about the constraints farmers face that limit their potential to develop tree resources within their farming system.

Experience with programmes to provide support to smallholder tree growing has consequently been variable. Many have failed to result in significant increases in tree resources. Others have, but often in unexpected ways or with unforeseen side-effects. The growing of trees as cash crops has, on occasion, attracted criticism on the grounds that it had negative impacts on food supplies, rural employment, and even on the environment. In some situations, farmers who initially adopted tree growing on a substantial scale have turned away from trees as a crop after a single production cycle.

At the same time, in other situations farmers were increasing the numbers of trees on their farmland, often on a large scale, without the benefit of any form of external encouragement or support. Contrary to the intuitive expectation that tree cover would diminish as pressures on the land heightened, agricultural intensification in some places was seen to be accompanied by higher densities of planted and managed trees.

This book has its origins in a research project at the Oxford Forestry Institute that was initiated in order to establish a better understanding of how tree management contributes to the strategies of farmers as they respond to pressures to intensify their use of land and other resources. It brings together the results of the work of a number of researchers who have been examining changes that have been taking place in different situations within eastern Africa and South Asia, two regions where tree growing figures prominently both in farmer practices and in government and donor programmes. It combines a number of studies that explore changes over time in a particular situation, or spatial variations in patterns of tree management across regions with different endowments and at different stages of development, with studies that examine the efficacy of trees and tree management in addressing different objectives, constraints and opportunities faced by farm households. Introductory and concluding chapters provide an overview of what emerges about the role of tree management in farmer strategies, and about the potential for policy and programme interventions.

The project was made possible by grants for the component research activities from the Rockefeller Foundation and the Ford Foundation, and by a core grant to the Oxford Forestry Institute from the Economic and Social Committee for Overseas Research of ODA. Most of the individual research studies were collaborative activities with institutions and individual researchers in the host countries, and would not have been possible without their unstinting support and participation. We are also deeply grateful to those of our colleagues from other institutions who developed analyses of their work specifically for use in the book.

A very large number of people contributed to the component studies, and we have tried to acknowledge their help in the reports on those studies. Here we would like to acknowledge our debt to those who helped shape and guide the project as a whole. The need for research on this subject was first suggested to us by Hans Binswanger and John Spears at the World Bank, and the discussions with them helped give it its initial shape. Gary Toenniessen at the Rockefeller Foundation, and Dianne Rocheleau and Eric Rusten in Nairobi and William Stewart and Mark Poffenberger in New Delhi within the Ford Foundation, were constant sources of encouragement and support over the period of several years we were working on this. In eastern Africa, Bjorn Lundgren and others at ICRAF provided invaluable assistance in a variety of ways. Within South Asia, Ravi Srivastava at the GIRI Institute and the University of Allahabad was a source of advice and help throughout the work in India, and Tushaar Shah and Vishwa Ballabh at the Institute for Rural Management at Anand (IRMA) organised and hosted a workshop that enabled us to benefit from an exchange of views with other researchers in the region. Finally, we would wish to acknowledge the support from Jeffery Burley and others at the Oxford Forestry Institute, and the benefit we have obtained from being able to test ideas contained in the book in the course of seminars and discussions with colleagues and graduate students there and elsewhere within Oxford University.

Oxford J. E. M. A.
June 1994 P. A. D.

Contents

Part III Factors influencing farmer decisions

Part IV Conclusions

List of Contributors

J. E. Michael Arnold,
Oxford Forestry Institute, University of Oxford, South Parks Road, Oxford OX1 3RB, United Kingdom.

Peter A. Dewees,
International Food Policy Research Institute, 1200 Seventeenth St., NW Washington, DC 20036, USA.

Michael R. Dove,
Program on Environment, East–West Center, 1777 East–West Road, Honolulu, Hawaii, 96848, USA.

Don A. Gilmour,
Coordinator, Forest Conservation Programme, International Union for the Conservation of Nature (IUCN), Rue Mauverney 28, CH-1196 Gland, Switzerland.

Narpat S. Jodha,
Head, Mountain Farming Systems Division, International Centre for Integrated Mountain Development, Kathmandu, Nepal.

N. C. Saxena,
Director, Lal Bahadur Shastri National Academy of Administration, Mussoorie-248179, India.

Sara J. Scherr,
International Food Policy Research Institute, 1200 Seventeenth St., NW Washington, DC 20036, USA.

Katherine Warner,
Regional Community Forestry Training Centre, Kasetsart University, Bangkok 10900, Thailand.

Part I
Overview

1 Framing the issues

J.E. Michael Arnold

1.1 Introduction

The presence of trees as part of contemporary farming systems has its origins
in two attributes of trees. One is their role in maintaining and restoring
the physical environment needed in order to sustain crop agriculture; most
notably through the restoration of soil nutrients and energy. The other is
the role various tree products play in helping sustain the rural household
economy. This includes products used directly by the household as food, fuel,
construction materials, etc.; inputs to agriculture such as fodder, mulch and
raw materials for making agricultural implements and storage structures; and
products or activities that provide household members with employment and
income. The presence or absence of trees may also have a role in securing or
maintaining rights of use or tenure, and certain trees or wooded areas can be
of cultural or religious importance.

This introductory chapter looks in general terms at how different patterns of
trees within farming systems have been shaped by these and other factors, and
examines the arguments that have influenced recent interventions in support
of tree growing by farmers. In doing so, many of the issues that are explored
in detail in later chapters are introduced, and the content of these subsequent
portions of the book is outlined.

1.2 Linkages between tree management and the farm household

Trees and smallholder agriculture

A substantial share of tropical smallholder agriculture is still based on a cycle
of clearing and burning the forest vegetation, a short period of cultivation
using the land and nutrients released in this way, and a longer period in which
the land is left fallow to allow the nutrient and energy capital stored in the tree
vegetation to be built up again. During clearance and cultivation care is taken
to preserve the root and seed stocks necessary to ensure the regrowth during
the fallow period. Tree species are encouraged which accelerate or enhance
the recycling of nutrients, or which produce outputs of value for subsistence
or income-generating purposes (Warner 1991).

Much of the share of tropical agriculture that has evolved to more intensive permanent use of land still reflects its origins in such shifting cultivation practices by retaining a tree component. As systems of settled agriculture have evolved, some have incorporated trees in order to perpetuate some of the nutrient restoring functions trees earlier performed in the fallow period of the cropping cycle. Leguminous and other nitrogen-fixing species may be intercropped in fields; or planted in off-field niches in order to provide green mulch to be cut and carried to the fields.

Having trees *in situ* on the farm may also provide some of the other physical benefits previously provided by the surrounding forest (shade from the sun, protection against torrential rain or soil loss due to surface water run-off, or protection against the desiccating effects of wind). The prevalence of windbreaks on plains, bands or patches of trees on hill slopes, and trees growing over shade-seeking crops such as cocoa, coffee or cardoman reflect these direct physical benefits to be obtained from incorporating trees in appropriate patterns within farm landscapes.

Trees play these roles even in the most intensively cultivated systems. Thus the tree components of the highly productive multi-storied multiple species home gardens of the wet tropics help maintain soil nutrients and structure, and create a micro-environment within which other plant and animal components can thrive, as well as contributing directly to the farm output through the production of fruits and other subsistence and commercial products.

Forest products and the rural household economy[1]

In many developing country agricultural systems, the farm household system depends on tree and other forest products for inputs that are often critical to the functioning of the system. These supplement other food, fuel and fodder supplies and income flows, fill in seasonal shortfalls of food and income, provide seasonally crucial agricultural inputs, and help reduce risk and lessen the impact of droughts and other emergencies.

For the majority of rural people, fruits and other forest foods add variety to diets, improve palatability, and provide essential vitamins, minerals, protein and calories. The quantities consumed may not be great in comparison to the main food staples, but they often form an essential part of otherwise bland and nutritionally poor diets. Many forest fruits and leaves, for example, are good sources of vitamin A, shortage of which is a common cause of blindness in many developing countries. Forest foods are also valued as snack foods, commonly eaten, especially by children, while working in fields, while herding and while gathering fuelwood.* Fuelwood is frequently the main source of energy used for cooking, which is often necessary in order to make foods

---

* The term 'fuelwood' is used throughout the book to denote wood burned as such, and 'woodfuel' for the total of wood used as fuelwood and as charcoal.

digestible and remove parasites, and for processing foods such as fish, seeds, and oil, that are generally smoked, dried or cooked in order to extend the food supply, or to provide a source of cash income. For those with limited resources, forest products often provide one of the few income earning options.

Products from forest and farm trees are most extensively used to supplement other resource and income flows during particular seasons in the year. Many agricultural communities suffer from seasonal food shortages during the time of year when stored food supplies have dwindled and new crops are only just beginning. In many arid regions, trees provide an important source of dry season fodder.

Forest and farm tree products are also valued during the peak agricultural labour period, when less time is available for cooking and people consume more snack foods. Home gardens are widely designed to make use of variations in the timing of the harvest of different component tree crops, in order to supply foods and saleable produce during the period between harvests of staple crops. Another important feature of such gardens, and other systems incorporating trees, is that work on the latter can often be undertaken during the slack season, thus helping to even out the peaks and troughs in the demand for farm labour (Ninez 1984).

A great many forest-based employment opportunities are also seasonal. The seasonality of some activities is dictated by the availability of the product or raw material, while in other instances it is determined by the demands of other activities such as agriculture—more time being available for gathering, processing and selling during the slack months. In other cases, the activities may be linked to seasonally induced cash needs such as loan payments or school fees. As the markets for many locally processed forest products are dependent on rural people's purchasing power, they too are tied to the cyclic nature of agricultural incomes.

A perhaps less obvious, but very important, role of forest-based activities is as an economic and environmental buffer—a source of income, fodder and even food that people can fall back on in hard times, or when other options fail. There is a wide range of forest resources, such as roots, tubers, rhizomes and nuts, that are eaten in periods of famine. More widely, the use and sale of gathered and processed forest products tends to increase as agricultural production declines, or when people need income to cover expenses of funerals, medical treatment or other contingencies (Chambers and Leach 1987).

The decline in off-farm tree resources

In most parts of the developing world, rural households have historically obtained most of the complementary inputs of fodder, fuel, green mulch, food and saleable commodities that are often critical to the continued functioning of their agricultural systems from nearby areas of forest, woodland or scrubland that were used as common property. However, nearly everywhere these resources, and their management and use systems, have been progressively eroded and undermined as a result of a long period of political, economic and physical changes.

State assertion of control first over the forest resource and then over the land has widely reduced access and rights of usage. At best people were left with usufruct rights, application of which was subject to the whim of the State and its officials. In recent times the reduction in availability of common property resources has nearly everywhere been massively accelerated. Privatization and encroachment, as well as government appropriation, have been the main processes taking resources out of common use. Increasing pressures on what is left have frequently led to its progressive degradation.

Concurrently, traditional methods of access control, usufruct allocation, and conflict resolution have widely become ineffective or have disappeared, undermined by political, economic and social changes within the village and nation. Increasing population pressure and in-migration of outsiders, greater commercialization of the products of the resource, and technological changes that encourage alternative uses of the land, have all contributed to increased differentiation within communities that reduces communal cohesion and uniformity of interest in the management of communal resources. With the progressive transfer of responsibility for resource management decisions to the central State, many common pool resources are no longer managed in any meaningful sense of that term by those who use them. Much usage is now of an unregulated 'open access' nature that accelerates the process of overuse and degradation (Jodha 1990; Poffenberger 1990; Arnold and Stewart 1991; Shepherd 1992; Messerschmidt 1993).

Shifts to private tree resources

As off-farm resources disappear or are degraded, farmers nearly everywhere have tended to shift the production of selected forest outputs of value to them on to their own land by protecting, planting and managing appropriate tree species. In recent times the process of adding trees to farming systems in some areas has been accelerated or transformed by expanding markets for fuelwood and other tree products, and the consequent emergence of the growing of trees as a cash crop.

The patterns of trees on farm that emerge vary greatly with the agroecological, economic and other factors bearing on a particular situation. The ecological constraints of arid areas, for example, dictate choices of land use, and of cropping and livestock management practices, that shape the situations within which tree growing can be located. Thus, water and site limitations rule out the intensive vertically structured vegetation patterns of home gardens (Cook and Grut 1989). The prevalence of free range livestock as part of the agricultural systems of dry lands restricts tree growing to locations that can be protected, such as homestead areas, or to tree species which are not browsed. The intensive crop agriculture possible where water is abundant is likely to discourage retention of trees within fields, causing tree growing to be relocated to boundaries and other niches not used for crops.

Despite this variation, tree growing is likely to evolve through a number of definable common stages (e.g. Food and Agriculture Organization of the United Nations (FAO) 1985; Raintree and Warner 1986). Where forest cover is locally abundant and population densities are low, tree management exists, but is usually passive. The offtake of tree-based products is usually offset by natural regeneration and tree growth. However, fruit trees are frequently planted even at this stage, and during land clearing farmers may choose to leave a few valuable trees on the farm. As population pressures increase, they may respond by leaving more trees during land clearance and by more intensively managing the remaining tree cover. Common tree management strategies such as coppicing, pollarding, and pruning, for instance, result in much higher total production than would have been possible if trees had been felled entirely.

As tree resources become increasingly scarce, farmers may take measures to stimulate tree regeneration; for instance, by protecting, transplanting, and cultivating naturally germinating seedlings. Fallow lands management of this type often favours the regeneration of particular trees, and is widely practised in both the wet and dry tropics. The leguminous species *Acacia senegal*, a valuable producer of gum arabic and other outputs in dry areas of Africa, and various pioneer species valued as building timbers in the high forest zone of West Africa, are instances.

Planting and farming of trees progressively emerges as a more intensive form of tree management when population and other pressures have reduced the farmer's access to woodland resources, or to introduce desired species that are not present in the existing tree stock, or to rearrange the tree resource within the farm landscape in response to the demands of other farm activities—such as ploughing, grazing of livestock, or burning to improve pasture. Trees may also be introduced at this stage to maintain some of the physical benefits associated with tree cover—shade trees for example, or trees such as *Acacia albida* that have beneficial effects on the yields of adjacent crops. As tree products become increasingly traded, trees may also become attractive to farmers as a cash crop.

As agricultural land use intensifies, the organization and sophistication of tree management strategies is likely to increase as well. It is at this stage that intensively managed home gardens evolve. With their layered vertical structure of trees, shrubs and ground-cover, they can provide more productive use of land and other farm resources than any alternative system. The woody perennials within the system contribute to nutrient recycling and

Table 1.1 Patterns of planted trees on farms

- **Trees on non-arable or fallow land**. This type of lower intensity management of naturally regenerated trees is likely to occur in more extensive farming and grazing systems.

- **Trees grown in homestead areas**. This often emerges even when there is still plentiful tree cover, to introduce fruit and other valued species. Where protection against livestock or burning is difficult, the homestead area can be the only niche where trees can be grown.

- **Tree growing along boundaries and in other interstitial sites**. Found where trees need to be separated from crops in areas of intensive land use, or where trees are the dominant means of boundary demarcation, or where lines of trees serve a protective purpose (e.g. windbreaks and contour planting).

- **Intercropping on arable land**. Generally takes the form of trees scattered, or in clumps or rows (alley-cropping), as part of sometimes complex agricultural crop production. Occurs where trees provide benefits to agricultural crops through shade, shelter or soil improvement, or intercropping is mutually beneficial to both trees and crops because of shared water, soil, nutrient, and light resources. In its most highly developed forms, as in multi-storied multiple species *home gardens*, tree/crop mixtures can represent important components of the overall farm system.

- **Monocropping on arable land** (farm woodlots). This is usually associated with the growing of trees to produce cash crops, such as poles, pulpwood, bark or for fruits such as cashew nuts, and is most commonly found in the more advanced market-oriented agricultural areas. Tree crops are also employed as a low cost means of using poor sites, or to maintain land as extensively managed fallow.

soil protection, yield produce that supplements outputs from other parts of the farm system, and can help spread farm work, outputs and income more evenly throughout the year.

The main forms in which planted trees are incorporated into agricultural landscapes are summarized in Table 1.1. The manner in which their occurrence and importance varies across different situations is explored for the eight-country region of eastern Africa in Chapter 5 (see also Warner 1993).

1.3 The debate about strategies to stimulate tree growing

Although knowledge about the tree component of agricultural landscapes was very limited, the heightened interest in it that began to emerge in the mid-1970s was not initially focused on learning more about the role it played. Instead, the debate about tree growing strategies that developed was heavily influenced by a number of more narrowly focused interests in encouraging trees within farming systems. These related to concerns about deforestation, environmental degradation and declining agricultural productivity, and meeting the fuelwood and other basic needs of the rural poor. Each of these perspectives provides some valid insights into one or more possible roles that trees can play in rural household livelihood strategies, in particular circumstances. But if pursued in isolation from each other, and without recognition of the broader framework within which farm trees are located, they are unlikely to provide a satisfactory picture of why farmers do or do not grow trees, for the reasons outlined in the two sections that follow.

Tree growing, deforestation and environmental stability

Much of the early impetus for intervention to stimulate tree growing by farmers stemmed from concern about deforestation. Growing household demands both for wood and for land were perceived as progressively reducing wood stocks, and denuding the land of tree cover that performed essential protective and regulatory functions. Promotion of tree growing on farms was seen to be necessary in order to create new wood stocks where they were readily accessible to the main body of users, thereby reducing pressure on remaining forests, and to re-establish a protective tree cover in environmentally fragile landscapes (e.g. Government of India (GOI) 1976; Bene *et al.* 1977; World Bank 1978).

This focus provided only a partial, and not necessarily very useful, perspective on the relationship of trees to farmer livelihood systems. In concentrating on the decline in forest and woodland cover as rural demands increase and more land has to be set aside for production of food, cash crops and livestock, it failed to recognize the extent to which the density of planted trees was often increasing, or the characteristics of these new on-farm resources.

The patterns of tree stocks and tree cover that emerge within farming systems are quite different from those to be found in natural forests. Tree resources created on farms serve particular quite narrowly defined purposes—production of fruit or fodder, shelter from the wind, boundary demarcation, etc. Farmers establish shelterbelts on windswept drylands, and plant trees along terrace risers or field boundaries in hilly land, in order to reduce soil erosion due to wind and water at the microlevel of the field or farm. But they are very unlikely to establish the large contiguous areas of tree cover needed, for example, to influence hydrological flows in mountain areas, or to create the species diversity found in a forest. Even a farm woodlot is not a functioning ecosystem in the sense that a forest is.

Thus, while tree growing by farmers may be, indirectly or directly, a response to the depletion of tree stocks due to deforestation, and can create additional supplies of wood and other forest products, it does not recreate forests. Trees in farming systems are therefore more usefully seen not as part of the forest resource, but in the context of farm household livelihood needs and strategies.

Tree growing, basic needs and poverty alleviation

A powerful second, rural welfare, perspective focused on the importance of tree resources in meeting people's fuel and other basic needs. Mobilizing farm households to grow more trees was identified as the most effective way for the rural poor to avert or reverse shortages of fuelwood and other essential tree products, from their own resources (e.g. Eckholm 1975, 1979; FAO 1978). As the diversity of goods and services derived from trees became better appreciated, a wider potential was postulated for 'agroforestry'—as a tool for resource-poor farmers in stabilizing and improving their farm system (e.g. Lundgren 1982; Nair 1984). Tree crops could help the poor to increase output and generate income, and secure a greater degree of self-sufficiency, with low inputs of capital and labour.

The weakness of the 'needs' approach was a failure to relate its concerns to farmers' other objectives, and to alternative uses of their resources. The growing of trees always involves some cost in terms of land, labour and capital invested. The produce of trees therefore has a real value to the farm household. As is explored in detail in Chapter 7, there will nearly always be alternative fuel supplies that are less costly for those for whom one or more of these factors of production are scarce resources, and there will nearly always be uses to which these tree products can be put that have higher value than as fuel.

Nor are people's needs static. Some of the changes that accompany, or contribute to, depletion of common pool tree resources, may alter demand for the tree products previously supplied from them. Improved access to markets, for instance, could make available fertilizer in place of green mulch, and create outlets for livestock products that cause livestock management to shift towards pasture crops and stall-feeding rather than grazing. Irrigation of dry land, to take another example, is likely to reduce the need for draught animals, and hence for fodder, and is also likely to create new and more productive sources of the latter than could be provided by fodder trees. Alternatives may become available that present a lower opportunity cost to the farmer than creating supplies of tree products—hence the widespread use of dung and crop residues in place of fuelwood. Other economic options available to the farm household—off the farm as well as on it—may offer a better use of its resources than adding or intensifying tree management.

Nor does all intensification of forest product and land use lend itself to increased incorporation of trees into agriculture. Some of the changes in agricultural land use may be such as to result in the elimination of trees from farming systems rather than their retention or establishment. Prominent among such pressures are competition with crops for light, water and nutrients on intensively used cropland; new agricultural techniques, such as the introduction of tractors which would be impeded in their operation by the presence of trees; broader land use practices such as burning of pasture and grazing of livestock which would make it impossible to protect trees; changes in legislation or customary practices concerning control of tree-bearing lands which inhibit private tree growing; and reduction in the rotational cycle to the point at which desirable trees are no longer able to re-establish. Alternatively, physical or environmental conditions such as poor soil or climatic conditions, inadequate water retention, poor nutrient balances, or saline conditions, may make the cultivation of trees impossible, or unprofitable as a smallholder activity.

The creation of new tree stocks on farm land in order to reproduce flows of forest products that were earlier drawn on to meet particular needs of poor households may therefore be neither efficient nor appropriate. Moreover, much poverty is associated with landlessness. Where the household does not have access to land, private tree growing evidently cannot have a significant role in alleviating poverty. The range of situations in which tree growing is a viable option for a poor farmer with access to a small amount of land is also limited by their need for annual rather than periodic income, and by the priority they attach to ensuring household food security.

To pursue tree growing by farm households as primarily a poverty alleviation measure can therefore be as counterproductive as trying to pattern it so as to combat deforestation—as is witnessed by the lack of success of most of the large stereotyped target-driven programmes and projects that have been put in place with this as their primary objective. It is necessary to identify those issues that tree growing can address, and those that it cannot, and these will vary greatly from situation to situation. Where trees do serve a purpose in terms of the household objectives, they will often contribute to meeting distributional, and environmental, concerns as well. But the perspective in assessing their role should be primarily that of their relevance to the goals of the farmer rather than the objectives of the environmentalist, the forester or the national planner.

1.4 The approach adopted in this book

Trees and the dynamics of rural change

The limitation of the partial, and static, analysis of tree product use and household demand reflected in the 'deforestation' and 'needs' approaches to analysis of the place of farm trees was that they tended to obscure the dynamics of farmers' economic responses to changes in demand and supply, and to scarcity and abundance. Solutions to real or apparent scarcities of tree products were framed in terms of tree growing options without adequate recognition of the alternatives, or of the adaptations to scarcity already practised by farmers.

A more holistic and dynamic framework for understanding such change is provided by the concept that households use the resources available to them to pursue a strategy that includes food security, social security, and risk management, as well as income generation, among their objectives (e.g. Chambers 1983). Over time, both objectives and resource availability will change, necessitating adaptations in the strategy the household pursues. The choices farmers make between different courses of action will also be influenced by changes in socio-economic conditions over time that alter the costs and returns they face (e.g. Ruthenberg 1980).

Using these concepts of 'livelihood security' and 'induced innovation' as a starting point,[2] tree planting can be explained as being one or more of four categories of response to dynamic change:

- to maintain supplies of tree products as production from off-farm tree stocks declines due to deforestation or loss of access;
- to meet growing demands for tree products as populations grow, new uses for tree outputs emerge, or external markets develop;
- to help maintain agricultural productivity in face of declining soil quality or increasing damage from exposure to sun, wind or water runoff;
- to contribute to risk reduction and risk management in face of needs to secure rights of tenure and use, to even out peaks and troughs in seasonal flows of produce and income, and in seasonal demands on labour, or to provide a reserve of biomass products and capital available for use as a buffer in times of stress or emergency.

The attractiveness of tree-based options in addressing these changes is likely to alter over time, as the relative prices and availability of the farmer's resources of land, labour, capital, and of other inputs and outputs, change. In most farm systems trees are present for a combination of more than one of the above reasons. Tree components of home gardens and compound farms, for example, typically contribute to all of the above.

The studies reported on in the rest of the book explore change within such a framework. The body of the book is divided into two parts. In the first, three chapters explore trends in tree growing over time in different situations within South Asia, and are followed by a fourth chapter that examines variations in spatial patterns of tree growing across eastern Africa. In the second part, the chapters address the factors influencing farmer decisions, looking first at decisions within predominantly subsistence strategies, and then at decisions relating to the growing of trees predominantly for the market.

Much of the content of the book is based on findings from a number of original case studies of different situations—high potential agricultural situations in north India (Uttar Pradesh) and central Kenya (Murang'a), dryland agriculture in western India (Rajasthan) and Pakistan, and predominantly subsistence agriculture in the Middle Hills of Nepal and western Kenya (Siaya and South Nyanza). In the Murang'a and Uttar Pradesh cases trees were being grown as cash crops, in the other cases the trees were meeting a variety of farm household objectives. In Uttar Pradesh and Pakistan the tree growing was largely a response to government 'farm forestry' programmes, in the western Kenya case tree growing was evolving with the support of extension services provided by a non-governmental organization NGO, and in the Nepal, western Rajasthan and Murang'a cases the changes were taking place spontaneously, without any overt external intervention directed at tree growing.

In each of the case study situations, and in the regional patterns examined in eastern Africa, the density of planted trees had increased during the period studied. However, in western Rajasthan the contribution of planted trees was very limited compared with the role that tree products from natural woody vegetation had played previously, in central Kenya there were shifts out of as well as into trees, and in Uttar Pradesh a large-scale reduction in tree growing occurred towards the end of the period.

Trends in farmer tree growing

The four chapters in this part of the book set out to explore changes in tree management over time in a range of situations with contrasting physical, socio-economic and institutional situations. Changes in tree cover are identified and examined, and are related to the changes in the broader framework within which they are located.

The first situation that is examined, by Don Gilmour in Chapter 2, is in the relatively isolated middle hills zone of Nepal. Comparison of aerial photo cover from 1964 and 1988 for selected plots in two districts in central Nepal showed that the density of trees on some categories of farm lands had increased substantially, despite increasing population pressures on the land, and the widespread view that tree cover has been decreasing. Field investigations in the same areas suggested that farmers pursue a strategy of natural regeneration and planting first on stream beds and banks and other uncultivated land, then on the walls of rain-fed terraces and then on the walls of irrigated terraces. Over time the density of planted trees in the total increases. Other evidence suggests that reduction in labour availability, increased access to markets, changes in fodder needs as livestock management practices evolve, and changes in adjacent public forest resources, may be factors influencing this increase in private tree management.

In Chapter 3, Narpat Jodha examines change in an area that has been exposed to much more change in recent times. This is the arid region of western Rajasthan (India), where people have historically based their livelihood systems on production of grain in association with nitrogen-fixing trees, on livestock management, and on retaining a substantial part of the lands as common property to ensure a reserve of biomass products for use in low rainfall years. The chapter records and examines the changes in these biomass-centred strategies over a period of more than two decades, in the face of land reform, heightened population pressures on the land, government programmes that have reduced the need for collective sharing, and progressive commercialization of agriculture in response to growing access to markets. During this period rapid depletion of the area of common lands, and overexploitation of the resource that remained, forced greater reliance on private tree management, but the adoption of tractor cultivation hindered growth in the latter.

In Chapter 4, recent changes in similar systems in nearby areas in Pakistan are examined by Michael Dove against the perspective of longer term historical changes. The chapter also looks at the political context of control of tree resources, and the impact this has on government interventions related to private tree growing.

In the last chapter in this part of the book, Chapter 5 by Katherine Warner, the focus shifts from examination of change over time in a particular situation to exploration of the variations in the patterns of tree growing across a whole region, covering a large part of eastern Africa. The differences observed are related both to the different endowments of different agroecological zones and, through cross-sectional comparisons, to factors such as agriculture and livestock management practices, rural dependence on forest products and availability of supplies from existing tree stocks, and commoditization of tree products and size of markets. Other factors looked at include tenure and rights, customary attitudes and practices, and government policies and interventions. An overall pattern emerges of increased planting as natural vegetation decreases, as land uses evolve within which trees can be protected from livestock and fire, and as access to markets, availability of inputs and the limited profitability of alternatives favour the adoption of trees as a farm crop.

Factors influencing farmer decisions

In the four chapters of the second half of the book, the focus is on the main factors likely to influence farmer decisions about tree management. These include the availability, cost and allocation of land, labour, and capital; the impact of subsistence demand and market opportunities; and farmer attitudes to risk and the management of different forms of risk.

In Chapter 6, Sara Scherr explores farmer decisions in a predominantly subsistence situation, within a framework that links such decisions to the concepts of livelihood strategy and induced innovation. This is applied to an area in western Kenya where on-farm tree planting and management have become progressively more intensive with the transition to permanent cropping, the disappearance of communal tree resources, and the rise of local cash markets for wood fuel, poles, seedlings and fruit. Farmers in this area employ a large and growing number of different tree species and management practices under conditions of increasing land scarcity, predominantly to obtain critical consumption goods which would otherwise have to be purchased, to protect food security in the face of declining crop yields, and to diversify their sources of cash income.

In Chapter 7, Peter Dewees explores farmer responses to tree product scarcity, using the case of woodfuel. The chapter shows how external perceptions of growing shortages of fuelwood, and of the options for responding by growing more trees differ markedly from those of the rural people concerned. Because the problem is fundamentally quite different from the way it has generally been perceived by governments and aid agencies, the responses of the latter have in large measure been inappropriate.

In Chapters 8 and 9, Peter Dewees and N.C. Saxena turn to tree growing predominantly for the market. Chapter 8 examines the conditions under which the market provides an effective incentive to farm level tree growing. The chapter compares the urban market for charcoal in Sudan, where farmer grown wood cannot compete with low cost supplies from natural woodland, the development of sustainable production of woodlots in central Kenya in response to market signals, and the widespread adoption of production of wood for markets by farmers in parts of north India, and their subsequent withdrawal as overproduction swamped the market and brought about sharp price falls.

In Chapter 9, they examine the impact of availability and allocation of land and labour on tree growing decisions, drawing on the case studies from central Kenya and north India. The growing of trees in certain situations is found to reflect its characteristics as an extensive land use requiring relatively low inputs of labour, and capital, but one that can reduce annual flows of food or income within the farm. In the Kenya case it was found that woodlots are more likely to be established as households age and as labour becomes scarce, and that woodlot clearance takes place when labour is available to cultivate more lucrative but intensive crops such as tea and coffee. In India eucalypt planting was taken up more by wealthier farmers who had more land, had diversified sources of incomes, and faced shortages of labour and problems of supervision.

In the final chapter, Michael Arnold presents a number of conclusions that emerge from the book. These focus on the need to be able to relate tree management to the dynamics of rural change, the balance to be struck between meeting household needs and responding to market opportunities, and policy and research implications of the findings.

Notes

1. This section draws on the work of J. Falconer contained in Falconer and Arnold (1989).
2. The basis of these concepts of 'livelihood security' and 'induced innovation' is

discussed further in Chapter 6, with particular dimensions of such farmer strategies being explored in detail in that chapter and in Chapters 7–9.

References

Arnold, J.E.M. and Stewart, W.C. (1991). *Common property resource management in India*, Tropical Forestry Paper No. 24. Oxford Forestry Institute, Oxford.

Beer, J.H. de, and McDermott, M.J. (1989). *The economic value of non-timber forest products in southeast Asia*. Netherlands Committee for IUCN, Amsterdam.

Bene, J.G., Beall, H.W. and Côté A. (1977). *Trees, food and people: land management in the tropics*, Report IDRC 084e. International Development Research Centre, Ottawa.

Chambers, R. (1983). *Rural development: putting the last first*. Longman, London.

Chambers, R. and Leach, M. (1987). *Trees to meet contingencies: savings and security for the rural poor*, IDS Discussion Paper No. 228. Institute of Development Studies at the University of Sussex, Brighton.

Cook, C.C. and Grut, M. (1989). *Agroforestry in sub-saharan Africa: a farmer's perspective*, Technical Paper No. 42. World Bank, Washington, DC.

Eckholm, E. (1975). *The other energy crisis: firewood*, Worldwatch Paper No. 1. Worldwatch Institute, Washington, DC.

Eckholm, E. (1979). *Planting for the future: forestry for human needs*, Worldwatch Paper No. 26. Worldwatch Institute, Washington, DC.

Falconer, J. and Arnold, J.E.M. (1989). *Household food security and forestry: an analysis of socioeconomic issues*, Community Forestry Note No. 1. FAO, Rome.

FAO (1978). *Forestry for local community development*, Forestry Paper No. 7. FAO, Rome.

FAO (1985). *Tree growing by rural people*, Forestry Paper No. 64. FAO, Rome.

GOI (1976). *Report of the National Commission on Agriculture, part IX: Forestry*. Ministry of Agriculture and Cooperation, Government of India, New Delhi.

Jodha, N.S. (1990). Rural common property resources: contributions and crisis. *Economic and Political Weekly, Quarterly Review of Agriculture*, 25 (26).

Lundgren, B. (1982). *The use of agroforestry to improve the productivity of converted tropical land*. Draft report for the Office of Technology Assessment, Congress of the United States, Washington.

Nair, P.K.R. (1984). *Soil productivity aspects of agroforestry*. International Council for Research in Agroforestry, Nairobi.

Ninez, V.K. (1984). *Household gardens: theoretical considerations on an old survival strategy*, Potatoes in Food Systems Research Series Report No. 1. International Potato Center, Lima, Peru.

Messerschmidt, D.A. (ed.) (1993). *Common forest resource management: annotated bibliography of Asia, Africa and Latin America*, Community Forestry Note No. 11. FAO, Rome.

Poffenberger, M. (ed.) (1990). *Keepers of the forest: land management alternatives in Southeast Asia*. Kumarian Press, Connecticut.

Raintree, J.B. and Warner, K. (1986). Agroforestry pathways for the intensification of shifting cultivation. *Agroforestry Systems*, 4, 39–54.

Ruthenberg, H. (1980). *Farming systems in the tropics*. Clarendon Press, Oxford.

Shepherd, G. (1992). *Managing Africa's tropical dry forests: a review of indigenous methods*, ODI Agricultural Occasional Paper No. 14. Overseas Development Institute, London.

Warner, K. (1991). *Shifting cultivators: local technical knowledge and natural resource management in the humid tropics*, Community Forestry Note No. 8. FAO, Rome.

Warner, K. (1993). *Patterns of farmer tree growing in eastern Africa: a socioeconomic analysis*, Tropical Forestry Paper No. 27. Oxford Forestry Institute and the International Centre for Research in Agroforestry, Oxford and Nairobi.

World Bank (1978). *Forestry: sector policy paper*. World Bank, Washington, DC.

Part II
Trends in farmer tree growing

2 Rearranging trees in the landscape in the Middle Hills of Nepal

Don A. Gilmour

2.1 Introduction

Nepal has an agriculturally based economy. Around 83 per cent of the population comprises rural dwellers who farm small areas of land at near-subsistence levels. The country is divided into a number of distinct physiographic zones which run parallel to the main Himalayan range (Fig. 2.1). The population of about 19 million is fairly evenly divided between the Middle Hills and Terai zones where most of the arable land is located. The Middle Hills zone is the heartland of Nepal and has been settled and cultivated for many centuries, originally by tribal groups, and from about AD 1000, by Hindu migrants from the south and west. Most of the Terai has been settled only during the past 40 years, and the process of agricultural expansion in that zone is still proceeding.

Recent studies (discussed in subsequent sections of this chapter) have indicated that the density of trees on some categories of farmland in the Middle Hills zone has increased substantially during recent decades. This increase in tree cover in the face of an ever-expanding rural population is in stark contrast to the scenario painted by Eckholm and others in the late 1970s (see for example Eckholm 1975, 1976; Seddon 1987; World Bank 1978). In this scenario it was postulated that Nepal was in the grip of an environmental crisis and that deforestation was one element of that crisis. This doom and gloom scenario has been dubbed by Ives and Messerli (1989) the 'Theory of Himalayan Environmental Degradation'. It was postulated that deforestation was taking place on such a scale that all accessible forest cover in the country would be lost by the year 2000. In fact, it has now been clearly demonstrated that deforestation in the hills region is not a recent phenomenon, and most clearance for agriculture was completed 80–100 years ago (Mahat *et al.* 1986). Aerial photographic evidence has also confirmed that there was no decrease in the area of forest in the Middle Hills between 1964 and 1978, at a time when it was widely perceived that deforestation in the hills was rampant.[1] However, in many cases the *density* of forest cover (as opposed to the area) was declining (His Majesty's Government of Nepal (HMGN) 1983).

The picture of an environmental crisis portrayed by Eckholm and others was a powerful influence in shaping the attitude of aid donors and Nepal's

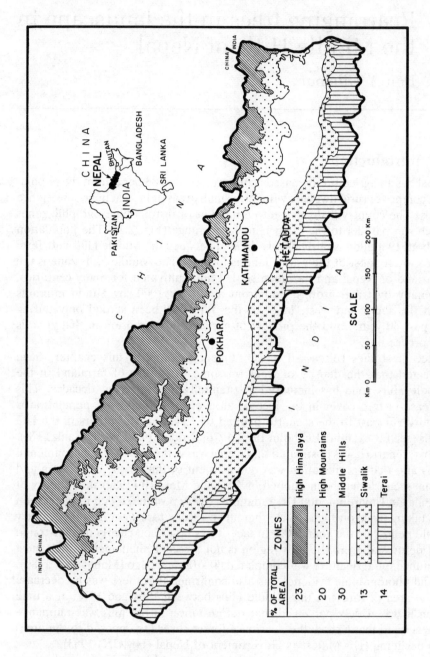

Fig. 2.1 Physiographic zones of Nepal.

ZONES	% OF TOTAL AREA
High Himalaya	23
High Mountains	20
Middle Hills	30
Siwalik	13
Terai	14

bureaucrats alike towards the identification of problems and the delivery of development aid. It was perceived that Nepali hill farmers had been forced into a situation through which they were destroying the resources upon which they depended for survival. This applied particularly to trees and forests. It was further perceived that farmers lacked the knowledge to appreciate the dimensions of the deforestation problem or the ability to do anything to address these problems. The findings presented in this chapter challenge this view and describe some of the ways in which farmers have reacted positively to changes in the availability of tree products and the extent of their indigenous knowledge about trees and their management. The results of several detailed studies from districts which span the hills region from the west to eastern-central Nepal are used for illustration.

2.2 Agricultural practices, and uses for trees and forest products in the Middle Hills

The present farming system in the Middle Hills is based on grain cultivation, generally on terraced land, supplemented by livestock herding. Irrigated terraces (*khet*) are used for rice production in the monsoon and wheat during the winter. These are generally located on lower slopes which enables irrigation water to be channelled to them. Upland, rain-fed terraces (*bari*) are used predominantly for maize and millet growing, usually as relay crops. Cultivated land in the Middle Hills constitutes only 27 per cent of the total land area, with the remainder being made up of grassland (7 per cent), shrubland (9 per cent), forest (40 per cent), and non-cultivated inclusions (15 per cent).[2] The most valuable land for crop production is *khet* but, because of limitations in soil quality and water availability, it accounts for only 30 per cent of the total cultivated land in the hills. All of the agricultural land is intensively farmed.

Farmers are dependent on various energy inputs to maintain their near-subsistence life-style and much of this energy comes from tree and other forest products. While the use of inorganic fertilizers is increasing in the more accessible areas, most farmers still rely heavily on organic forms of fertilizer to maintain soil productively. One of the main reasons for keeping large animals in the hills is for the 'conversion' of leaves and other herbage to dung. Edible tree leaves make up a high proportion of the fodder fed to animals, particularly during the winter months when ground forage is in short supply. Large quantities of tree leaves (both green and dry), often mixed with agricultural residues such as rice or wheat straw, are used as bedding material in animal stalls. When mixed with dung and composted, this material becomes the principal fertilizer used in cultivated fields.

Even in areas where animal herds are grazed mainly in the forest, they are almost always stalled at night where they are fed on forage which is cut and carried from the forest. Similarly, animals which are stall-fed most of the time are occasionally open-range grazed. This is particularly necessary for oxen in order to maintain their muscle tone for draught work.

The extent to which tree leaves are used for fodder and bedding varies greatly depending on such factors as the size and type of herd, the extent of stall feeding, availability of household labour and access to suitable fodder trees. Rusten (1989), quoting the results of several studies from throughout the hills region, estimated that the total amount of fodder obtained from trees ranged from 13 to 90 per cent. He also collated data which indicated that between 20 and 80 per cent of tree fodder comes from privately owned trees. Although there is great variability in dependency on fodder trees, Robinson and Thompson (1989) emphasized that they 'often provide the only green fodder during the critical times of the dry season'.

Mahat *et al.* (1987) carried out a study in a farming community east of the Kathmandu Valley where farm animals were stall-fed for most of the year. They measured the amount of biomass used as fuel, fodder, and animal bedding material and found that fodder and animal bedding biomass was 7.5 times greater than fuel biomass. Most of this material was collected from adjacent forests (normally by women) and carried to the farms. While it is often perceived that fuelwood is the major forest product utilized by farmers, this study gives some indication of the importance of forests in providing fodder and bedding material as well as fuelwood for a typical farming area in central Nepal. Farmers devote a considerable amount of their time and energy to collecting fodder, of which tree leaves are a major component, particularly during the winter months.

Fuelwood is the major source of fuel, although, in areas where wood is in short supply, woody agricultural residues such as maize stalks and cobs are used extensively. As with estimates of fodder usage, estimates of fuelwood usage vary widely depending on factors such as altitude and access to fuelwood source. It is difficult to obtain meaningful averages of fuel use. Various studies have estimated consumption to range from 400 to 2000 kg of dry fuelwood per person annually (Metz 1989).

While tree leaves and fuelwood comprise the major tree products required to maintain the farming systems, timber is required for house construction and for manufacturing farm implements. In areas where livestock are kept in temporary animal shelters (*goths*) in field and forest, substantial numbers of poles are required for construction and maintenance of the *goths*.

There are many minor (in quantity) but none the less important tree and forest products which are extensively harvested and used in hill farming systems. These include utilitarian items (such as special woods for turning into carrying containers) as well as food (fruits and fungi). There is a whole culture which utilizes ayurvedic medicines, and much of the raw material for these medicines is derived from forest areas.

This discussion has painted a picture which depicts trees as being not just an adjunct to farming, but rather an *essential* ingredient in subsistence farming

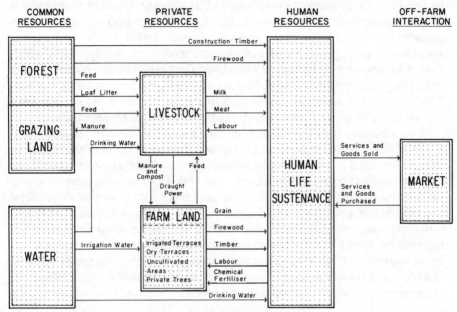

Fig. 2.2 Integration of trees and forest products into farming systems in the Middle Hills of Nepal.

systems. Without access to tree and other forest products, farming systems in the hills would not survive. A simplified model of the energy flows between forests, trees, and other components of the farming system in the Middle Hills of Nepal is shown in Fig. 2.2.

2.3 Tree and forest management in the Middle Hills

Management of tree resources on public land

Farmers generally have access to tree and forest products on both their private land and on public 'common' land.[3] Land-rich farmers may be able to grow sufficient trees on their own land to satisfy most of their needs but land-poor farmers cannot. They must rely on public lands to supply the bulk of their needs. While almost all of the public lands are legally National Forest, much is under some form of indigenous management system. Reviews of the literature and field studies during recent years have revealed the widespread presence in the Middle Hills of indigenous management systems in public forests (Campbell *et al.* 1987; Fisher 1989; Joshi 1990; Tamang 1990).

Such systems are 'extra-legal' and are generally not known or acknowledged by the Forest Department bureaucracy. The simplest forms of management do little more than define access rights (for instance, who can go to which areas to collect what products, and who cannot; Fisher 1989). Such simple systems are quite widespread and do not necessarily involve rules which have a silvicultural or biological objective. They may not be concerned with protecting regeneration, or with limiting usage within the defined user group. This explains the common observation that forests under local management are declining in quality (tree density and/or amount of regeneration).

More advanced forms of indigenous management involve an explicit recognition of the need to conserve or to improve the composition of the forest. In these cases, local rules are framed accordingly and those who transgress these rules are subject to agreed sanctions. The shift from management based on the definition of access rights only, to management with a specific silvicultural objective is probably the result of a community perception of 'need'. In simple terms the community perceives that the forests are degrading and something needs to be done about it. This is a rather simplistic view of a complex set of social interactions, but it has been argued (Gilmour 1990) that this perception of need is a necessary precondition to stimulate the initiation and modification of indigenous management of common forests.

Management of tree resources on private land

One of the surprising findings of recent investigations about tree cover on private land is the widespread extent of tree planting and protection of natural regeneration which is taking place. This finding is surprising because of the pervasive effect of the 'Theory of Himalayan Environmental Degradation' which has resulted in the common view held by many government officials and aid workers that substantial environmental degradation (particularly, the loss of tree cover) is proceeding (Gilmour 1988).

Anecdotal information provided by farmers in two districts in central Nepal, indicated a substantial increase in tree cover during the past 10–20 years on *bari* land and on non-cultivated patches within the agricultural complex (Gilmour 1988). Observations indicated that farmers normally kept the main cultivated part of their *bari* terraces free of trees, but allowed trees to regenerate naturally (or planted them) on the edges and terrace risers. This initial finding was validated and quantified by comparison of the tree cover on *bari* land in 1964 (using aerial photographs) and in 1988 (using on-ground surveys) along a transect in the same two districts (A.S. Carter and Gilmour 1989). For the purposes of the study, trees were enumerated only when they exceeded certain criteria such that they could be counted on the aerial photographs, and shrubs and bushes were excluded. Four sites were selected at different elevations along the transect and 15 plots of *bari* land, with an average area of about 1 ha, were selected randomly at each site. Table 2.1 shows the results of the survey.

Table 2.1 Changes in tree densities between 1964 and 1988

Study site	Tree density (number per ha), by year		Increase between 1964 and 1988	Per cent increase between 1964 and 1988
	1964	1988		
Nala	27	126	99	367
Devitar	46	192	146	317
Sangachok	63	262	199	316
Saping	126	614	488	387
Average	65	298	233	357

Notes: Calculated for trees over 2 m in height, at each of four study sites. Measurements were taken at an average of 15 plots per study site. Source: A.S. Carter and Gilmour (1989).

The average increase in density from 65 to 298 trees /ha is the *net* increase, as farmers also cut and removed trees to satisfy their needs for fuelwood and construction timber. The more than fourfold increase in tree density on *bari* land shown in Table 2.1 is a clear indication of the substantial private forestation effort by farmers, involving both the protection of naturally regenerated seedlings and the planting of selected high value species. It became clear during the study that the changes had not come about by chance, but were a result of conscious choice and action by farmers, which governed both the number of trees protected and the type of management afforded to each tree.

Instances were seen where previously bare terrace edges had been planted during the past 1–3 years with *Prunus cerasoides* seedlings dug out of adjacent forest areas. The trees were evenly spaced and well tended. It was common to see saplings that had blown over propped up on stakes. Saplings on path edges were often protected from browsing by being wrapped in woven cane baskets. As well as actively cultivating and protecting trees, farmers are also actively utilizing them in a very sensible fashion. Species suitable for construction, such as *Alnus nepalensis*, often have the side branches removed (and used for fuel) so that a long stem suitable for building purposes develops. Such a tree shape casts minimum shade onto the surrounding agricultural land. Clearly, farmers are very tree conscious and actively propagate, protect, and utilise desirable species. The present farming practices can best be described as sound agroforestry: farmers are integrating trees and agricultural crops on their *bari* land (A.S. Carter and Gilmour 1989).

The species which occur most commonly seem to be those that can be propagated easily and which regenerate naturally. Species of very little or no value are largely excluded, but the species assemblage is less dominated by the generally recognized high value fodder species, such as *Ficus* sp. and *Litsea monopetala*, than might be expected.

Temporal changes in the spatial distribution of trees in the farming landscape

A perusal of the various aerial photograph runs together with current observations of the same locations, allowed hypotheses to be developed to explain the changing patterns of tree cover in the farming complex. It was postulated (A.S. Carter and Gilmour 1989) that the forestation process on private land has been evolving through a series of stages as depicted in Table 2.2. Stream beds and banks seem to be important in this forestation process. Aerial photographs showed that, in many of the areas investigated, stream beds and banks were totally devoid of trees in 1964. Observations in 1989 of the same areas indicated that the same stream beds and banks were frequently densely tree covered. They are often the first areas to become tree covered, generally by natural regeneration, and thus act as germplasm sources for the spread of trees on to the *bari*.

A.S. Carter and Gilmour attempted to quantify the extent of increases in tree cover on the major categories of private land in the two districts where they carried out their study. In these two districts of Sindhu Palchok and Kabhre Palanchok there are about 62 000 ha of *bari* land (Land Resources Mapping Project (LRMP) 1986a). It could be assumed that a small proportion of this *bari* may not have shown any substantial increase in tree cover because of its proximity to existing sources of forest products, and because there is consequently no need for the farmers to propagate trees on their private land. This may exclude 10 per cent of the *bari* leaving about 55 800 ha. If the average increase over the 24-year period of 233 trees /ha applies to this area of *bari*, it means that the farmers in the two districts have succeeded in increasing the net number of trees on their *bari* land by about 13 million. This is equivalent to a forest area of 6500 ha (assuming a planting density of 1600 stems /ha).

The land-use estimates by LRMP (1986b) indicate that there is an area of 47 500 ha of non-cultivated land included within the agricultural land in these two districts. A limited sample by Gilmour (1988) indicated a tree density of about 600 trees /ha on these non-cultivated inclusions. Oral history reports suggest that a threefold increase in tree cover has also occurred on this land during the past 20 years (Gilmour 1988). This has resulted in the addition of a further 17 million trees to the landscape, equivalent to a forest area of about 9500 ha (assuming that 90 per cent of the available land has been influenced). Thus, the total number of trees added by farmers to both *bari* land and to the non-cultivated segments of farmland in the two districts during the past

Table 2.2 Stages in the evolution of tree cover, central Nepal

Stage 1	Stream beds and banks bare of trees.
	Non-cultivated land bare of trees.
	Terrace wall of *bari* bare of trees.
	Terrace wall of *khet* bare of trees.
Stage 2	Stream beds and banks with discontinuous tree cover.
	Non-cultivated land largely bare of trees.
	Terrace walls of *bari* largely bare of trees.
	Khet edges bare of trees.
Stage 3	Stream beds and banks become more densely tree covered.
	Non-cultivated land becomes lightly tree covered.
	Terrace walls of *bari* developing sparse tree cover.
	Terrace walls of *khet* bare of trees.
Stage 4	Stream beds and banks with continuous tree cover.
	Non-cultivated land becomes more heavily tree covered.
	Terrace walls of *bari* become well tree covered, but with few trees per clump of woody vegetation.
	Terrace walls of *khet* are for the most part bare of trees.
Stage 5	Stream beds and banks covered with dense, continuous tree cover.
	Non-cultivated land is tree covered.
	Terrace walls of *bari* are tree covered, with many trees per clump. Some *bari* land removed from agriculture, as trees take over formerly cultivated land.
	Isolated trees are found of terrace walls of some *khet*.

Source: A.S. Carter and Gilmour (1989).

24 years could be as high as 30 million, equivalent to a forest area of about 16 000 ha (A.S. Carter and Gilmour 1989).

While this detailed study was confined to two districts in central Nepal, extensive enquiries by the author and colleagues across other parts of the Middle Hills have indicated that similar changes are occurring in many other areas.

2.4 Factors influencing farmer tree management decisions

Land holding size and its relationship to tree growing

The median farm size in the hills is less than 0.5 ha (Seddon 1987). Most farmers own some *khet*, *bari* and non-cultivated patches. Farmers tend to concentrate tree planting and protection of naturally regenerated seedlings on selected portions of their land. Table 2.3 shows the results of a survey which determined the number and density of trees on different parts of the farms for three villages east of Kathmandu.

The data in this table indicate that most of trees are found on non-cultivated

Table 2.3 Location and number of trees on private land in study area

Land-use category	Area under type of land use (ha)			Number of trees[a]		Density (trees per ha)
	Mean	Mini- mum	Maxi- mum	Mean	Maxi- mum	
Irrigated terraces (*khet*)	0.66	0.00	30.00	27	305	41
Rain-fed terraces (*bari*)	0.93	0.05	3.00	73	425	84
Around houses	–	–	–	19	78	–
Marginal land	0.17	0.00	0.55	80	1000	624
Stream banks	0.10	0.00	1.00	141	1125	1371
Total	1.86	0.05	6.00	340	2125	183

Notes: [a] In all land-use categories, the number of trees planted varied from zero to the maximum.
Field measurements were taken on 20 private landholdings in the villages of Acharyagoan, Timilsinagon, and Chhapgaon in Phulbari Panchayat of Khabhre Palanchok District.
Source: Gilmour (1988).

inclusions and along the edges of the *bari*. The largest area of land available for tree growing in most farms is the *bari*, and in one study, an average of 357 trees /ha on *bari* land was recorded in a sample of 60 sites totalling 54 ha (A.S. Carter and Gilmour 1989). In the four major sites sampled in this study, four species were found to account for more than half of the total number of species at each site. However, the four most common species generally varied from site to site (with the exception of *Schima wallichii*, *Litsea monopetala*, and *Prunus cerasoides*; Table 2.4). As the sample covered *bari* land only, the total number of species recorded was relatively low (from 29 at the highest elevation site of 1774 m to 60 at the lowest elevation site of 1050 m).

In another study, on farmland at a high elevation site in eastern central Nepal (1700m–2200 m), E.J. Carter (1991) also noted that the tree population of 101 species was dominated by four species: *Ficus nemoralis*, *Alnus nepalensis*, *Brassiopsis hainla* and *Schima wallichii*. At a lower elevation site (1000–1700 m) in the same general area she recorded a total of 111 different species on farmland. The population was dominated by one species, *Buddleia asiatica*, which made up 25 per cent of the total number of trees. Rusten (1989) also noted that a small number of species dominated the total population of 127 species at his sample site in western Nepal (at an elevation of 1800-2300 m).

What is evident from these three studies is that a small number of species generally dominates the assemblage, but that the dominant species vary from place to place even where the elevation is roughly the same. A.S. Carter and Gilmour (1989) found in their study that the low elevation sites had the greatest representation of high value fodder species. For example, at the

Table 2.4 Frequency of most commonly found tree species and their uses, at four study sites

Site	Altitude (m)	Total number of recorded species	Most common species	Frequency as a per cent of all trees	Usage rating[a]			
					Fodder	Fuel	Fruit	Construction
Nalla	1774	29	Prunus cerasoides	39	3	2	4	1
			Betula alnoides	15	3	2		1
			Alnus nepalensis	14	1	1		2
			Ficus nemoralis	9	1		2	
Devitar	1405	46	Prunus cerasoides	17	3	2	4	1
			Litsea monopetala	15	1			
			Psidium guajaya	12			1	
			Pinus roxburghii	10		2		1
Sangachok	1050	60	Bauhinia yariegata	16	1		2	
			Litsea monopetala	16	1			
			Schima wallichii	14	3	2		1
			Ficus semicordata	13	1		2	
Saping	1108	39	Litsea monopetala	29	1			
			Schima wallichii	23	3	2		1
			Bridelia retusa	10	1		3	
			Psidium guajaya	9	1		1	

Notes: [a] Represents the most valued usages. Rating system adapted from Howland and Howland (1984).
Source: A.S. Carter and Gilmour (1989).

highest elevation site in their sample (1774 m) the three most common species (all non-fodder) made up 68 per cent of the total. By contrast, at the lowest elevation site (1050 m), three high value fodder species made up 45 per cent of the total.

The lower elevation sites were also the ones with the greatest total density of trees on *bari* land. Oral history information collected from elderly farmers indicated that most of these areas had been bare of trees in their living memory. However, 1964 aerial photographs showed that even at that time, the forestation process was well advanced on these lower elevation sites. In the early stages of the process, natural regeneration is probably responsible for most of the new growth as it could take many years for high value fodder trees, many of which would be artificially propagated, to become a significant component of the assemblage. After the tree density increases, the micro-environment within the developing tree covered areas also becomes more conducive to the establishment, both natural and artificial, of some of the higher value species. This is rather speculative, but it is consistent with the difficulties experienced in attempting to establish many high value fodder species in open plantations. Observations on incursions of natural regeneration of the high value *Litsea monopetala* under 12 year old *Pinus* plantations (Gilmour *et al.* 1990) also support the suggestion that reasonable protection and site amelioration is needed before some of the high value fodder species can be easily established.

Rusten (1989) reported from western Nepal that as land size decreased below 1.5 ha farmers concentrated their tree planting on fodder producing species, with 47, 60 and 65 per cent of all trees being fodder trees for the land size categories of 1.0–1.49, 0.5–0.99 and 0.0–0.49 ha, respectively.

One observation which is common to all studies is the paucity of nitrogen-fixing trees among the species commonly reported as growing on private land. *Alnus nepalensis* is the major exception, but this species is only common on moist sites. This has raised the question about the trade-offs between agricultural production and tree growth (A.S. Carter and Gilmour 1989). Almost invariably farmers acknowledge that the presence of trees on *bari* edges adversely affects crop production but they see no alternative. As they frequently say, '. . . without trees we cannot survive.' There is probably substantial scope for introducing nitrogen-fixing trees into the agroforestry system so that there are fewer adverse effects on agricultural crops. Legumes, such as locally available *Albizzia* sp., are not widely used and exotics such as *Leucaena* have proved suitable only in specialized localities at lower elevations (A.S. Carter and Gilmour 1989).

In the same study it was noted that sites with a high overall density of tree clumps also had large numbers of trees per clump. It seems that, as the forestation process proceeds, the density of trees in each clump steadily increases. This is perfectly logical. When a terrace edge is first planted or regenerates naturally, there will be only isolated trees but, with time, clumps develop. Even where artificial establishment is used to initiate the forestation process, natural regeneration probably becomes dominant particularly for species which regenerate from root suckers. Thus, as private forests age and mature, tree density and species diversity increases and a larger number of small seedlings and saplings become evident. This explains the observation that the greatest absolute increases in tree cover between 1964 and 1988 occurred on sites that had the highest tree densities in 1964 (A.S. Carter and Gilmour 1989).

In the area that he studied in western Nepal, Rusten (1989) noted that, with the exception of the 'near-landless farmers' (who owned and farmed land as small in area as 0.05 ha), there was relative uniformity of tree density across landholding categories (Table 2.5), with an average of 458 trees /ha.

Malla and Fisher (1987) in a similar study in central Nepal found that farms larger than 1.5 ha in area had twice the tree density (225 trees /ha) of smaller farms (104 trees /ha). These studies are not conclusive in terms of the relationship between farm size and tree density probably because of the importance of other influences. None the less it is clear that all farmers, large and small, have recognized the economic advantages of having trees on their farms.

Small farmers, however, have insufficient trees for all of their needs. Discussions with land-poor farmers have shown clearly that they see limits to the numbers of trees that they can tolerate close to their crops because of competition. Several poor farmers said that it was preferable to obtain off-farm labouring work and to use part of the money to buy trees from their wealthy neighbours periodically rather than to plant more trees on their small areas of *bari* terraces. Such land-poor farmers are also the ones who need to rely on access to common forests to satisfy many of their forest product needs. They are thus further disadvantaged compared with their wealthier neighbours who do not need to expend as much labour in the collection of forest products or have to pay cash for any of those products.

Table 2.5 Average tree density by size of landholding

Landholding size (ha)	Trees per ha	Trees per household	Number of households
0.00–0.49	526	141	30
0.50–0.99	440	293	13
1.00–1.49	215	242	5
1.50 or greater	363	605	6

Source: Rusten (1989).

Establishment, management, and use of fodder trees

Investigations in Nepal during the past decade have revealed that there is a substantial and sophisticated knowledge system about trees, their products, and their management. Rusten (1989) described a system of tree fodder evaluation from a village in western Nepal which others have suggested is in wide usage throughout the country (Robinson 1990). In this system, tree fodders are evaluated by farmers as being either *chiso* or *obano*. These terms relate to the characteristics of the fodder when they are fed to animals.

For example, feeding of *obano* fodder tends to:

- lead to the production of good, firm, and relatively dry dung without causing constipation;
- improve the general health of livestock;
- cause livestock to gain weight;
- be eaten well and satisfy the animal's appetite;
- contribute to the production of milk and clarified butter.

By contrast, feeding *chiso* fodder tends to;

- cause animals to produce watery dung;
- sometimes cause weight loss;
- weaken animals and cause them to lose their appetite and possibly cause a blockage in the stomachs or throats;
- not satisfy the animals' appetite;
- not increase the production of milk suitable for making clarified butter.

Rusten found that some fodders can move along the scale from *chiso* to *obano* depending on the time of year and the growth stage of the plant.[4] Rusten described how farmers balance the feed to their livestock depending on their production goals at any particular time, and argues that the observable seasonal variation in the use of particular fodders reflected farmers' assessment of changes in the 'ripeness' of the fodder. E.P. Rusten (personal communication) indicated that users would harvest only those trees, or parts of a tree, on which the leaves had matured to the point at which they possessed the fodder qualities needed at that time, returning to harvest from other parts of the tree as they ripened.

While this *chiso/obano* system of evaluation is not related to modern scientific plant classification, it does give a sense of order to nature's apparent disorder. Most importantly, it gives farmers a common frame of reference to discuss issues of vital importance to their well being. E.J. Carter (1991) also described other criteria by which fodder values are rated.

In addition to knowledge about the properties of different tree fodders, many farmers have substantial knowledge about different propagation techniques for important fodder trees. Robinson and K.C. (1990) carried out a fodder tree survey in several districts west and south of Kathmandu at

Table 2.6 Species propagated vegetatively by farmer, and method used

Species	Number of respondents	Direct planting of cuttings	Air layering	Cuttings first rooted in moist soil
Brassiopsis sp.	8	●	●	
Erythrina sp.	1	●		
Ficus glaberrima	6	●	●	
Ficus lacor	15	●	●	
Garuga pinnata	8	●	●	●
Saurauia nepalensis	1	●		
Bakhari[a]	2	●		
Dumre[a]	1			●
Sami[a]	2		●	
Kursimlo[a]	1	●		

Notes: [a]Scientific name not known.
Source: Robinson and K.C. (1990).

elevations between 850 and 1500 m. They found that many farmers possessed substantial knowledge about the vegetative propagation of trees. Table 2.6 shows the species which surveyed farmers were propagating vegetatively and the methods they were using.

The survey also found that, of 27 species mentioned by the respondents as being used for fodder, 10 species are propagated by vegetative means. Farmers also showed considerable knowledge about the conditions which are important for the successful propagation of vegetative material, such as the optimum diameter and length of the cutting, the age of the branch and its stage of growth, and soil and plant management, to bring about successful rooting.

Farmers also experiment with different methods of tree establishment. Rusten (1989) reported an innovative technique for propagating an important fodder species, *Ficus nemoralis*, in western Nepal:

Because of the palatability of this species, grazing animals make it very difficult to propagate, especially on public lands. To overcome this problem, farmers in Salija use *Neolitsea umbrosa*, a small bushy tree that grazing animals ignore, as a nurse plant for *F. nemoralis*. From field observations and according to farmers who use this technique, companion planted *F. nemoralis* grows more quickly than trees grown without *N. umbrosa*. (Rusten 1989)

Clearly, farmers have a sophisticated level of knowledge about matters relating to fodder trees and their management. This is not to infer that all farmers possess all of this knowledge, but within the farming community as a whole, substantial knowledge does exist. Hobley (1990), for instance, in a study of several ethnic groups in central Nepal, noted that tree planting and tree ownership were considered to fall within the male domain, and women exhibited little knowledge of these activities. It is evident from extensive discussions with farmers that, while some high value species are propagated, most trees on private land are the outcome of managed natural regeneration.

Household demands and their impacts on tree cultivation and management

Trees, like other resources at a farmer's disposal, are managed according to the requirements of the household. These requirements vary from region to region and, within one region, from farm to farm, depending on a whole host of factors. The perception of many who are involved in intervening in rural development is that all farmers, everywhere, should plant more trees. This simplistic view is neither sensible nor appropriate, yet many projects are designed with just such a basic rationale.

It has been postulated (Gilmour 1990) that the extent to which shortages of forest products have developed is a major influence on the extent and complexity of indigenous management systems on both private and communal land. In other words, there is a continuum ranging from an abundance, to a severe scarcity, of forest resources, and the attitude of people toward the use and management of the forest will vary depending on where they are along the continuum.

The likely response of farmers with regard to tree planting and protection under three different levels of scarcity of forest produce is illustrated in Fig. 2.3. While this model may be somewhat simplistic it does go a long way toward explaining the different attitudes of farmers in different parts of the hills. In simple terms, those who live close to an accessible and available source of tree and forest products have little interest in cluttering up their agricultural land with trees which will reduce their yield of agricultural crops. Farmers at the other end of the availability continuum have responded by establishing large numbers of trees on their private land (either by direct planting or by protecting naturally regenerated seedlings) and by becoming involved in communal efforts to manage the common forest land.

Changes in the rural economy

Changes in tree growing also reflect broader changes in farming practice and in the rural economy in parts of the Middle Hills which are becoming more affected by, and integrated into, the market economy. In a study of changes

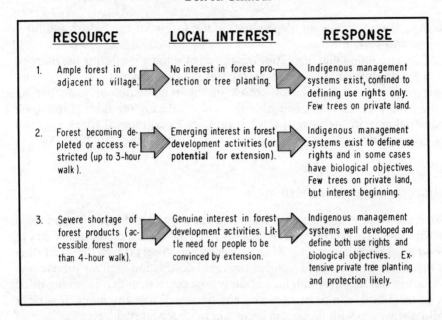

Fig. 2.3 Accessibility of forest resources, and probable responses of villagers to tree scarcity and abundance (from Gilmour 1990).

to farming systems in an area east of the Kathmandu Valley, Y.B. Malla (personal communication) noted that, because of increasing market influences, traditional farming practices are changing rapidly. In many places the availability of off-farm employment has resulted in a shortage of farm labour. This, in turn, has caused changes in the traditional patterns of collection of forest products. Less labour is now available to make the onerous daily trek into the forest to collect leaf material for animal bedding for compost, and fodder for feeding the animals. The result is that more trees are cultivated on land close to the house to make collection easier. Many farmers in the area now claim to be self-sufficient in forest products and no longer need to rely on common forests. This applies particularly to land-rich farmers.

Y.B. Malla (personal communication) reported that farmers in some areas reduce their need for organic manures because of the ready availability of chemical fertilizers in this area, and because of the increased access to cash through the market economy. This reduces the need to maintain large numbers of animals to produce manure. In addition, the development of the Kathmandu milk market has resulted in a marked change in the number and type of large animals kept. It is now common for an average family to maintain one or two stall-fed buffalo rather than 20 or more open range grazed cows. Buffalo are more efficient milk producers than cows and respond better to stall feeding. These changes have substantially reduced pressures on forests, because fewer animals are dependent on the forest for the fodder and litter required to sustain them.

Malla also found that increasing quantities of wood are entering the market, with 46 wood-using industries now registered in the District and many more operating without registration. Most of these industries only started up during the past 10 years. Much of the wood entering the market, for both industrial and private uses, is *utis* (*Alnus nepalensis*) most of which is grown on private land.

Changes in government policy

Government regulations which restricted the cutting and sale of wood from private land were relaxed several years ago and this facilitated the legal entry of wood into the cash economy. Many government officers feared that the lifting of these restrictions would result in the rapid destruction of many private and public forests. However, this has certainly not occurred, and trees are regarded by farmers as a crop to be nurtured and harvested like any other. It is clear from discussions with farmers in areas where strong market influences prevail that trees are seen as a crop which can add substantially to their well-being as they can be held like money in the bank, gathering interest, until needed for an emergency. Hobley (1990) reported similar findings among farmers east of Kathmandu. Nevertheless, government forest officers still exercise substantial control over trees on private land, and government policy regarding the sale and movement of trees from private land is far from being clear and straightforward. This leads to confusion on the part of forest officers and to mistrust by the farmers. Government practice constrains the achievement of its own policy objective of encouraging farmers to manage their own tree crops for their own benefit.

Concurrent, and sometimes related, changes have been taking place on public forest lands. Work carried out by the LRMP indicated that the tree density of much of this resource was declining steadily (HMGN 1986). However, this trend is not universal. As was noted above, in some areas changes in livestock husbandry have substantially reduced pressures on the forest. More widely, there has been an emergence of knowledge about indigenous forest management systems (Campbell *et al.* 1987; Fisher 1989; Joshi 1990; Tamang 1990). In many areas, interviews with farmers have indicated that where such indigenous management systems incorporate protective elements and are functioning effectively, the density of the forest is increasing steadily, and conditions are considered to be better now than they were 20 years ago. In a study in a densely populated 14 000 ha catchment east of Kathmandu, Gilmour and Nurse (1991) found that the per cent tree crown cover of the common land had increased significantly, from 61 to 81 per cent, between 1972 and 1989 (Fig. 2.4). The area studied was one where there had been a

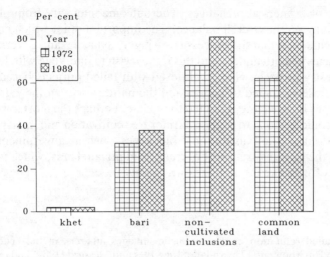

Fig. 2.4 Changes in tree crown cover between 1972 and 1989 (from Gilmour and Nurse 1991).

substantial reduction in the size of livestock herds (thus reducing the pressure on the forests to provide fodder) and at the same time the emergence of effective indigenous systems of forest protection.

2.5 Conclusions

There was a period, possibly covering many centuries, during which tree cover over most of the hill regions was reduced steadily as the population expanded its agricultural base. Government taxation policies during the nineteenth century encouraged the conversion of forest to agricultural land. Various accounts (Macfarlane 1976; Gilmour 1988; Hobley 1990) have suggested that shifting cultivation was widespread, on what has become *bari* land, 50–100 years ago probably because labour was a limiting factor. Also, up to that time, livestock husbandry was a much more dominant part of the farming economy than is the case in some areas today. During relatively recent times there has been a move to sedentary, labour-intensive, terraced agriculture as populations have expanded rapidly. This has been accompanied by a steady decline in the importance of livestock husbandry.

The combination of economic, demographic, social, political, and resource availability factors has determined the outcome of how farmers manage their private farmland and their adjacent common land. As far as trees are concerned, it is clear that farmers have been involved in a process of 'rearranging' the trees on the landscape under their immediate control to suit their needs at any particular time. These needs change from time to time and the response of farmers also changes. The outcome may vary from place to place depending on the overall balance. The important point which emerges from this discussion is that the processes are highly dynamic, and governments and aid agencies need to understand this if they wish to intervene with forestry and agroforestry projects. Nepal is littered with failed projects because the designers themselves failed to understand the nature (or even the existence) of the processes in which they planned to meddle. Perhaps the most important thing which could be done to support farmer tree cultivation and management would be to remove any institutional blockages, such as government rules constraining the sale and transport of trees from private lands, which prevent farmers from freely exercising their own talents.

Notes

1. For a detailed refutation of the simplistic linkages inherent in the 'Theory of Himalayan Environmental Degradation' see Ives and Messerli (1989) and Gilmour and Fisher (1991).
2. Non-cultivated inclusions refer to the small patches on non-cultivated land within the general agricultural complex. They are included with cultivated land in the land-use mapping described by HMGN (1986).
3. For a detailed discussion on the distinctions between common property and open access resources and the way in which common property regimes operate in forest land in the Middle Hills of Nepal, see Fisher (1989) and Gilmour and Fisher (1991).
4. Others, such as E.J. Carter (1991), have been unable to corroborate these findings in other areas.

References

Campbell, J.G., Shrestha, R.J. and Euphrat, F. (1987). Socio-economic factors in traditional forest use and management. Preliminary results from a study of community forest management in Nepal. *Banko Janakari*, **1**, 45–54.

Carter, E.J. (1991). Private tree cultivation in part of Dolakha District, Nepal: a case study. Paper presented at the Institute of Rural Management workshop on *socio-economic aspects of tree growing by* Farmers in South Asia at Anand, India, March, 1991.

Carter, A.S. and Gilmour, D.A. (1989). Tree cover increases on private farm land in central Nepal. *Mountain Research and Development*, **9**, 381–91.

Eckholm, E.P. (1975). The deterioration of mountain environments. *Science*, **189**, 764–70.

Eckholm, E.P. (1976). *Losing ground*. W.W. Norton, New York.

Fisher, R.J. (1989). *Indigenous systems of common property forest management in Nepal*, Environment and Policy Institute Working Paper No. 18. East-West Center, Honolulu.

Gilmour, D.A. (1988). Not seeing the trees for the forest: a re-appraisal of the deforestation crisis in two hills districts of Nepal. *Mountain Research and Development*, **8**, 343–50.

Gilmour, D.A. (1990). Resource availability and indigenous forest management systems in Nepal. *Society and Natural Resources*, **3**, 145–58.

Gilmour, D.A. and Fisher, R.J. (1991). *Villagers, forests and foresters*. Sahayogi Press, Kathmandu.

Gilmour, D.A. and Nurse, M. (1991). *Farmer initiatives in increasing tree cover in central Nepal*. Paper presented at the Institute of Rural Management workshop on socio-economic aspects of tree growing by farmers in South Asia at Anand, India, March, 1991.

Gilmour, D.A., King, G.C., Applegate, G.B. and Mohns, B. (1990). Silviculture of plantation forests in central Nepal to maximise community benefits. *Forest Ecology and Management*, **32**, 173–86.

HMGN (1983). *The forests of Nepal: a study of historical trends and projections to 2000*, Report No. 4/2/200783/1/1. Ministry of Water Resources (Water and Energy Commission), Kathmandu.

HMGN (1986). *Land use in Nepal: a summary of the Land Resources Mapping Project results*, Report No. 4/1/310386/1/1 Seq. No. 225. Ministry of Water Resources, (Water and Energy Commission), Kathmandu.

Hobley, M.E.A. (1990). Social reality, social forestry: the case of two Nepalese panchayats. Unpublished PhD thesis. Canberra, Australian National University.

Howland, A.K. and Howland, P. (1984). *A dictionary of the common forest and farm plants of Nepal*. Forest Research and Information Centre, Department of Forestry, Kathmandu.

Ives, J.D. and Messerli, B. (1989). *The Himalayan dilemma: reconciling development and conservation*. United Nations University and Routledge, London and New York.

Joshi, A.L. (1990). *Resource management in community forestry: a literature review*. Centre for Economic Development and Administration, Tribhuvan University, Kathmandu.

LRMP (1986a). *Land utilisation report*, Appendices 2 and 3. LRMP, His Majesty's Government of Nepal, Kathmandu.

LRMP (1986b). *Land utilisation report*. LRMP, His Majesty's Government of Nepal, Kathmandu.

Macfarlane, A. (1976). *Resources and population: a study of the Gurungs of Nepal*. Cambridge University Press, London.

Mahat, T.B.S., Griffin, D.M. and Shepherd, K.R. (1986). Human impact on some forests of the Middle Hills of Nepal. 1. Forestry in the context of the traditional resources of the state. *Mountain Research and Development*, **6**, 223–32.

Mahat, T.B.S., Griffin, D.M. and Shepherd, K.R. (1987). Human impact on some forests of the Middle Hills of Nepal. 4. A detailed study in South East Sindhu Palchok and North East Kabhre Palanchok. *Mountain Research and Development*, **7**, 111–34.

Malla, Y.B. and Fisher, R.J. (1987). *Planting trees on private farmland in Nepal. The equity aspect*. Discussion Paper. Nepal/Australia Forestry Project, Kathmandu.

Metz, J.J. (1989). The *goth* system of resource use at Chimkhola, Nepal. Unpublished PhD thesis. University of Wisconsin, Madison.

Robinson, P.J. (1990). Some results of the Dolakha private tree survey. In *Proceedings of the Third Working Group meeting on fodder trees, forest fodder and leaf litter*, Occasional Paper 2/90. pp. 40–8. Forest Department, Forest Research and Information Centre, Kathmandu.

Robinson, P.J. and K.C., S. (1990). Summary of results of a survey of the vegetation propagation of fodder trees by farmers. In *Proceedings of the third working group meeting on fodder trees, forest fodder and leaf litter*, Occasional Paper 2/90. pp. 35–9. Forest Department, Forest Research and Information Centre, Kathmandu.

Robinson, P.J. and Thompson, I.S. (1989). Research needs in fodder trees. In *Proceedings of a workshop on research needs in livestock production and animal health in Nepal*, pp. 70–85. HMG/MA/NARSC/ARPP, Kathmandu.

Rusten, E.P. (1989). An investigation of an indigenous knowledge system and management practices of tree fodder resources in the Middle Hills of central Nepal. Unpublished PhD thesis. Michigan State University.

Seddon, D. (1987). *Nepal, a state of poverty*. Vikas Publishing, New Delhi.

Tamang, D. (1990). *Indigenous forest management systems in Nepal: a review*, Report No. 12. Ministry of Agriculture and Winrock International Research Report Series. Kathmandu.

World Bank (1978). *Nepal: Community forestry development and training project*, Staff Appraisal Report. World Bank, Washington, DC.

3 Trends in tree management in arid land use in western Rajasthan

Narpat S. Jodha

3.1 Introduction

Farmers manage trees as an integral part of their farming system, within a given ecological and social setting. The pattern and intensity of management of trees in terms of resource allocation, protection, upkeep and usage depend on the importance of trees or tree-based biomass in the functioning of the system. The role of trees within it can alter over time, as people's perceptions, needs and decisions change. The evolution of local practices and strategies, which occurs through a process of informal interactions and experimentation, is also likely to be influenced by a variety of public interventions. The health and sustained productivity of the natural resource component of the land use system depends upon formal public policies being compatible with people's informal strategies for resource management.

The role of trees is likely to be particularly important in farming systems in environments characterized by high risk and low production, where nature's regenerative processes are slow and farming is best confined to extensive land uses based on the natural vegetation. The present chapter records changes in such a situation: the arid region of western Rajasthan, which covers 202 000 km^2 within the tropical arid zone of India. Severe physical and climatic constraints, and the consequent need for collective security, historically encouraged village communities to keep large parts of their lands as common property resources (CPRs). Trees and shrubs play an important role in people's production and usage strategies in this environment, both as CPRs and as a component of the farming system itself, which is based on the production of grain in association with nitrogen-fixing trees, and on livestock management.

The chapter reports on the findings from a series of studies over a period of two decades that provide information about changes in these biomass-centred strategies as institutional, demographic, and technological circumstances have evolved.[1] During this period the area experienced a series of political and administrative reforms that transformed control of land and traditional patron–client and leadership patterns, population growth that heightened pressures on the land, and progressive commercialization of agriculture in response to growing markets, expanded transport infrastructure and closer links with wider state and national situations.

The chapter is divided into five sections. The second section briefly describes the constraints on production in the arid region of western Rajasthan, and uses village and farm level evidence to describe the biomass-oriented strategies that farmers and pastoralists in the region have evolved in response. The third section records the decline and reduced effectiveness of traditional strategies under the pressure of changed institutional, demographic and technological circumstances. The fourth section discusses the ways people have adapted to these changes. A final discussion section considers the main factors influencing the further evolution of tree management, and implications for public policies and strategies.

3.2 Societal adaptations to the arid environment

Key constraints and potentials

The arid region of western Rajasthan consists of the 11 north-western districts of the state. The region is characterized by highly erodible sandy or sandy loam soils with low moisture retention, and scattered rocky patches in some areas, high growing season temperatures, desiccating winds, low and highly erratic rainfall, and deep and often saline ground water. After mapping the areas according to a land capability classification that had been developed in the region,[2] Jodha and Vyas (1969) reported that, unless transformed by irrigation, nearly 79 per cent of lands in the arid region are not suited to the high intensity of use involved in crop farming.

A number of important consequences of the region's natural resource base may be noted. First, the highly erodible and nutrient-poor soils and paucity of moisture that characterize much of the region favour activities like pasture-based livestock rearing rather than crop farming. Secondly, the low and unstable rainfall and short period of moisture availability make production of grain more uncertain than production of other biomass outputs from the same crops. Thirdly, the natural vegetation of trees, shrubs, bushes, perennial grasses (and even annual grasses at time), etc., are relatively less sensitive to length of wet period and fluctuations in rainfall than the domesticated crops. Maintenance of existing vegetation therefore imparts greater stability to production of biomass in arid areas.

Historically, people's responses to the above constraints, and potentials, have focused on those land use practices and cropping systems that enhanced quantity and certainty of crop and non-crop biomass availability, and that thereby helped sustain livelihood activities without unduly increasing the intensity of use of land. The traditional farming systems and land use practices discussed below demonstrate this.

Biomass strategies

Adaptations to the arid environment can be seen in the form of various practices, ranging from scattered settlement patterns and nomadism on the one hand, to folk agronomy and ethno-engineering[3] designed for conservation and security of water on the other (Jodha 1991). As trees and shrubs are integral parts of their biomass strategies, farmers' management of trees and changes can best be understood in terms of the traditional systems of the desert farmer that are designed to avoid or counteract scarcity and instability of biomass supplies. In what follows the focus is on the interrelated practices that farmers employ in pursuit of these biomass-centred production and usage strategies.

The key components of farmer and pastoralist strategies that contribute to this goal are indigenous agroforestry, crop–fallow rotation, common property resource management, folk agronomy emphasizing biomass production, collective security measures against biomass scarcity, crop–livestock based mixed farming, and supply-led adjustments in the use of biomass. Table 3.1 summarizes the contribution each makes to meeting different objectives of the farmer. Using the village and farm level evidence from the studies mentioned earlier, the extent and functioning of these tree-related practices are explored in more detail below.

Mixing extensive and intensive uses of land   The farmer in the arid areas tries to achieve the twin goals of higher biomass production (especially fodder), and lower intensity use of land (to avoid erosion), through a mix of several measures. Two of the measures employed are crop–fallow rotation and agroforestry practices involving indigenous bushes and trees like *khejri* (*Prosopis cineraria*; Mann and Saxena 1980).

1. Crop-fallow rotation: Data from villages in Jodhpur and Nagaur districts indicate that 29–38 per cent of their area of cropland was fallowed during the period 1963–65 (Table 3.2). This not only helped in rebuilding soil fertility but provided for grazing and collection of fodder, fuel and other non-crop material. Some of the land (called *bira* protected fallow plots) was fallowed for a prolonged period specifically to provide for collection of fodder. Fast growing shrubs, especially *ber* (*Ziziphus numularia*), play an important role in this system.

2. Indigenous agroforestry: In the study villages a number of trees and shrubs are maintained within the fields, especially *khejri* and *ber* bush colonies, as shelterbelts on field boundaries (Table 3.2). These constitute another important source of fodder, fuel and other material, while keeping part of the crop land under natural vegetation. *Khejri* trees are lopped for fodder and fuel every year after the harvest of rainy season crops (Mann and Saxena 1980). *Ber* bushes are cut back to ground level before planting the field crops. They regrow from stump sprouts interspersed among the crop, and are harvested again after the harvesting of the latter. Trees, bushes and shrubs grown as shelterbelts are not harvested, but camels and goats browse on them.

Table 3.1 Key features of biomass-orientated strategies of farmers in western Rajasthan

Components of strategy	Trees and shrubs play an important role	Biomass supply	Annual-perennial links	Extensive land use	Collective security	Adjustment to weather	Flexible demand-supply	Conversion of biomass into economic gains
Indigenous agroforestry	•	•	•	•		•	•	•
Common property resources management	•	•	•	•	•	•		
Folk agronomy								
Crop-fallow rotation	•	•		•		•		
Cultivar choice[a]		•		•		•		
Crop-livestock mixed farming	•			•	•	•	•	•
Supply-led management practice[b]	•						•	•
Seasonal migration					•	•	•	

Notes: [a] Crops with high stalk grain ratios, and high salvage potential.
[b] Extent of storage, processing, recycling according to relative scarcity/abundance of supplies.

Table 3.2 Indicators of changes in land use and biomass harvest in study villages in Jodhpur and Nagaur Districts

	Districts, by years			
	Jodhpur		Naguar	
Details	1963/65	1982/84	1963/65	1982/84
Landholding data				
Sample farms/holding (number)	38	38	43	43
Average landholding (ha)	9.3	7.8	8.4	7.1
Proportion of land holding (per cent):				
fallowed in total[a]	29	15	38	14
fallowed as *bira* for fodder harvest	11	4	30	14
planted to crops	71	85	62	86
Plot data				
Total plots monitored (number)	27	27	32	32
Khejri trees (number per ha)	20	26	18	29
Ber bush colonies (number per ha)	8	3	15	9
Plots with shelterbelts (number)	8	0	17	2
Number or years when sampled plots were harvested from 1972/73 to 1982/83				
for grain yield		6		5
for fodder from crops		8		7
for top feed from trees		10		10
for top feed from bushes		10		10
for fencing/fuel material		10		10

Notes: Data from two villages in each district.
[a] Unintended fallows due to rain failure are excluded.
Source: Jodha (1968, 1986).

The most striking fact revealed by the data on the use-history of the monitored plots, is the relative stability and dependability of biomass production from indigenous agroforestry. In contrast to the field crops, which are more affected by the occurrence of rain, fodder from *khejri* and from *ber* is available every year (Table 3.2). Hence they provide farmers with a measure of natural insurance. During drought years tree and shrub fodder fetches as high a price as food grains. Another stabilizing attribute is that the timing of harvest of edible pods and fodder from the perennials, and of the labour demands of the harvesting operations, fit conveniently into the crop cycle. An equally important feature of the leguminous trees and bushes retained in fields is the beneficial impact they can have on adjacent crops. During good rain years the crop yields around *khejri* trees and *ber* bush colonies are often higher than in other parts of the same plots. In addition to its positive impact on soil nutrient levels, *khejri* provides shade which reduces soil temperatures and conserves soil moisture, both problems for crop germination in sandy soils. Its canopy also protects the pearl millet (*Pennisetum americanum*) crop against desiccating winds (Michie 1986).

The emphasis on high biomass production and an extensive pattern of land use is also reflected in the choice of crop cultivars within folk agronomy systems. Crops like local *bajra* (pearl millet) and *jowar* (sorghum), with longer maturity period, indeterminate type (i.e. crops having recurrent flushes of flowering), and high stalk-to-grain ratios, are preferred. Similarly, high priority is given to crops which have high salvage potential (i.e. the possibility, in the event of mid-season rain failure, of harvesting fodder even if there is not a grain harvest).

Complementary use of perennials and annuals   In addition to their function in enhancing crop performance, the presence of perennials like *khejri* trees and *ber* bushes is of value to the farmer as a more stable source of biomass. Data collected during different rounds of fieldwork in the villages in Jodhpur and Nagaur (Table 3.3) revealed the following:

1. During years with poor rainfall *khejri* and *ber* top feeds play an important compensatory role in overall supplies of fodder. This is partly because they are less sensitive to fluctuations in rainfall, and partly due to greater efforts by the farmer to harness them during poor rain years.
2. During poor rainfall years there is also greater emphasis on collecting fuel, fencing and thatching material from perennials, on common land as well as farmland, for own use as well as for sale. With little coming from crops, the perennials bear the brunt of providing supplies during drought periods.
3. During years with good rainfall crop by-products contribute the larger share of total biomass production, and biomass from agroforestry components is not fully harnessed. Tree and bush top feeds remain unharvested and in many cases pearl millet stalks as well (partly due to labour shortages and partly due to complacency).

The full set of data, collected in each of 8 years spread over the period 1963–64 to 1983–84, also makes clear the greater stability of biomass supplies from perennials. It is this that induces dryland farmers to give high priority to the protection of trees and shrubs on their croplands.

Complementarity of common and private property resources   The emphasis on biomass production and extensive use of land is not confined to the household and plot level situation. It is also pursued at the community level, by keeping part of the village land (often the sub-marginal lands) as village

Table 3.3 Differences in patterns of management, harvest, and use of fodder and fuel during years of scarcity and abundance in study villages in Jodhpur and Nagaur districts

| | Districts, by year | | | |
| | Jodhpur | | Naguar | |
Details	1963/64	1964/65	1972/73	1973/74
Rainfall (mm)	159	377	198	510
Biomass yield from different sources[a]				
Fodder (kg/ha) from:				
top feeds from *khejri* and *ber*	377	305	420	390
grass	85	780	400	615
crop by-products	62	2250	915	1240
Grain (kg/ha)	7	503	214	558
Fuel/fencing/thatching material (cart load) from:				
crop by-products	0	3	0	3
others	2	2	4	2
Per cent of sample plots in which biomass supplies were increased through[b]				
collection of weed as fodder	33	3	48	0
harvesting field borders	43	4	39	3
harvesting pre-mature crops	10	0	42	0
harvesting grain crops as fodder	31	0	42	0
premature harvesting of;				
ber bush for fodder, etc	58	0	70	2
khejri tree loppings for fodder and fuel	45	2	65	0
Per cent of sampled households undertaking the following measures[c]				
use stalk:				
after chaff cutting	98	14	100	36
without chaff cutting	2	86	0	64
leftover fodder of productive animals reused for feeding unproductive animals	54	5	78	12
leftover fodder (waste) mixed for making dung cakes	60	6	94	20
bajra stalk and *ber* left unharvested	0	7	0	12
fuel/fencing/thatching material collected from CPRs	48	16	52	20

Notes: Rainfall data was collected from nearest rain gauge station.

[a] Data from 20 sample households which could be covered in all rounds of fieldwork.

[b] Number of plots monitored was 60 and 88, respectively, in the villages in Jodhpur and Nagaur districts.

[c] Number of sample households covered by above information was 62 and 75, respectively, in two villages each in the districts of Jodhpur and Naguar.

Source: Field surveys reported in Jodha (1968, 1974, 1985a, 1986).

CPRs which every member of the community can use. Common property resources include village pastures, forest, wasteland, watershed drainage areas, and ponds (Jodha 1985a). Villagers, particularly the rural poor, depend very heavily on these for supplies of forage, cut fodder, fuel and other materials to supplement what they can produce or gather on their own lands.

This is illustrated by data from villages in Jodhpur, Nagaur, Jalore and Barmer districts (Jodha 1986). Though much reduced from their extent 30 years previously, CPRs still constituted 21–28 per cent of the total land resources available in these villages in 1982–84. The shares of various biomass products coming from these CPRs in the different villages during the same period was as follows:

- from 66 to 79 per cent of animal grazing;[4]
- about 55–65 per cent of fuel (including dung) collected by sample households;
- about 29–41 per cent of the collected fodder (non-crop products);
- about 59–68 per cent of fencing and thatching material (which is ultimately recycled as fuel).

Collective security against periodic scarcities   Several institutional arrangements involving trees and shrubs have been evolved by the desert people in order to mitigate the impacts of scarcity of fodder and other biomass in times of poor rainfall. The provision of village lands as CPRs is most important. Concern for collective needs and mutual help is also reflected in what may be described as seasonal CPRs, whereby during the post–harvest season fellow villagers are allowed free access to private lands for animal grazing, use of top feeds, food gathering, and collection of dung and other material.

Similarly, under the system of periodic closure of parts of village territories to animals, individuals' rights to graze even on their own lands are suspended. This closure of territory occurs in the spring season, beginning around the month of *chaitra* (late March–April), to permit regrowth and sprouting of trees, shrubs, bushes and perennial grasses. Contributing to charity feeding during periods of scarcity is another collective risk sharing system. Seasonal migration in groups, and cattle maintenance entrustment on mutually agreed terms (Jodha 1986), are other measures employed to help each other. Most importantly, there are several informal institutional arrangements in force, such as imposition of community sanctions, that are designed to regulate operation of collective security measures.

Mixed crop and livestock farming   Mixed farming based on crop production and rearing animals helps ensure balance between extensive and intensive uses of land, as it keeps some of the area under natural vegetation that is less sensitive to rainfall variability. Linkages between farming, forestry, pasture, and livestock also contribute to maintaining a diversified biomass component in the farm system. Marketable animals and animal products generate economic benefits from biomass-oriented activities. Livestock, being mobile, are also able to adjust to rain-induced spatial and temporal variations in biomass supplies. A livestock component therefore enables an aridland farmer to take fuller advantage of other components of his system, such as agroforestry, crop–fallow rotation, and CPRs.

Supply-led flexibility in biomass management   In the arid areas of western Rajasthan, farmers adapt their harvest and usage of biomass resources to the relative scarcity or abundance of supplies, as the latter fluctuate with rainfall. A comparison of biomass production and consumption practices in the study villages of Jodhpur and Nagaur districts during the years with poor and good rainfall illustrates how this is implemented in practice (Table 3.3). During poor rainfall years farmers were found to augment production and collection of biomass through a series of measures, ranging from collection of weed material as fodder or fuel to pre-mature harvesting of tree and shrub products. In addition, a number of measures to economize on use of biomass, such as chaff cutting and reuse of leftover fodder, were adopted during years of scarcity.

As most of these measures are inferior options, they are not practised during years of good rainfall. The proportion of the usable biomass that remains unused or wastefully used in such years represents the extent of slack or buffer resources present in the system.

3.3 Changes in traditional strategies

The biomass-oriented farmer strategies outlined above had evolved in the context of an environment characterized by low pressure on land, subsistence orientation of agriculture, and absence of rapid institutional and technological changes. However, all these conditions have altered in recent decades, putting pressure on traditional strategies (Jodha 1985a, b, 1990).

The strategies incorporated various arrangements to facilitate tree management in terms of three sets of interrelated activities:

1. allocation and protection of areas devoted to trees (and shrubs);
2. measures to maintain and enhance the productivity of trees;
3. patterns of usage, usage regulation and access to tree products.

However, as pressures on the resource base have grown, several of these arrangements have weakened and in the process have adversely affected farmers' tree management systems. These changes are summarized in Table 3.4, and are elaborated upon below.

Impacts of changes in CPRs on tree management

The earlier practice of protecting, upkeep, and regulating usage of CPRs was accompanied by protection and regulated use of trees and shrubs. As the area and quality of management of CPRs have declined over the last three to four decades, the protection, upkeep, and productivity of trees have also deteriorated. The component parts of this process can be described as follows.

Decline of area   The decline in the area of common lands in villages of the arid zone has been largely due to privatization of CPRs through governmental land distribution policies. Following the introduction of land reforms, the area of CPRs in the 11 districts comprising the arid zone as a whole has declined

Table 3.4 Changes in traditional biomass-focused strategies that have had adverse impacts on tree management

Components of strategies and changes therein	Tree management aspects affected		
	Area protection and usage regulation	Upkeep, productivity and maintenance of trees	Access to and usage of tree products
Common property resources (CPRs)			
Privatization of commons	•		•
Conversion into croplands	•		
Disruption of management systems		•	•
Physical degradation	•		
Reduced collective use of seasonal CPRs		•	•
Indigenous agroforestry			
Tractorization affecting population of woody perennials		•	
De-emphasis of collective sharing			•
Reduced landholdings and overextraction of woody perennials	•	•	
Folk agronomic systems			
Reduced extent of land fallowing	•		
Reduced collective sharing			•
Reduced annual–perennial links		•	
Public programmes to upgrade production and resources	•	•	•

from 11.3 million ha in 1951–52 to 8.4 million ha in 1981–82, reducing the proportion of the area in CPRs from 61 to 44 per cent. In the study villages, the decline in the proportion of village area remaining as CPRs over the 32 years from 1950 to 1982 ranged from 28 to 68 per cent.

More importantly, the land reforms did not involve any obligation on the part of the land recipients to use land according to its use capability. Consequently over 80 to 90 per cent of privatised common property lands were shifted from natural vegetation to annual cropping with low and unstable productivity (Jodha 1990).

Slackening of management   The introduction of land reforms was accompanied by certain institutional reforms, including democratic administration of villages through the system of village *panchayats* (elected village councils), and substitution of formal, legal measures in place of the informal arrangements that had guided the protection, development and use of CPRs. However, *panchayats* not only failed to enforce any management of CPRs, but their introduction led to discontinuation of practically all the practices conducive to protection, development and sustained use of CPRs. Grazing taxes, fees, levies and penalties are no longer enforced; measures to regulate CPR use such as rotation of grazing, periodical closure of parts of the common property area, and posting of watchmen, are not practised any more; and measures to raise revenue to pay for management of CPRs through auction of dung collection rights, top feeds and wood, and investment in maintaining the productivity of CPRs, have been discontinued (Jodha 1985a). In essence, overexploitation and underinvestment have become the key features of most communities' approach to CPRs (Jodha 1990).

Physical degradation   A consequence of shrinkage of common property areas, their overcrowding, and absence of upkeep as well as usage regulation, has been the physical degradation of CPRs. This is reflected in the reduced quantity and quality of tree, shrub, and other products available from the CPRs. The data from four CPR plots summarized in Table 3.5 indicates rapid decline in output of most products by the early 1980s compared with the late 1940s (Jodha 1985a).

Reduced access to seasonal CPRs   Factors such as increased pressure of population, slackening of social sanctions, villagers' reduced interest in collective sharing, and the changes associated with land distribution and the introduction of village *panchayats*, have also adversely affected the provision of seasonal CPRs. Private lands previously used as CPRs during the non-crop season are now rarely accessible to outsiders. This has also adversely affected seasonal migration of animals from one area to others during periods of local scarcities (Table 3.6).

Table 3.5 Decline in productivity of CPRs in four forest and grazing plots in a village of Nagaur District

Product	Plot 1 (6 ha)			Plot 2 (10 ha)		Plot 3 (12 ha)			Plot 4 (12 ha)	
	1945/47	1963/65	1982/84	1945/47	1963/65	1945/47	1963/65	1982/84	1945/47	1963/65
Timber (*babul* and *indok* trees)	12	3	0	11	1	3	0	–	17	0
Top feed from *khejri*	8	4	2	10	3	21	8	5	12	3
Top feed from *ber* bush	–	–	–	7	3	12	4	0	15	2
Fuelwood (*khejri, ber*, etc.)	8	2	1	5	2	18	6	3	21	4
Cut grass (*kared* and *dhaman*) perennials	13	3	2	18	4	27	9	2	21	0
Cut grass (*bharoot* and other annuals)	3	5	3	5	7	10	8	5	13	9
Dung collection	–	–	–	–	–	15	0	–	17	0
Gum (*babul* and *indok* trees)	40	0	–	10	0	–	–	–	–	–

Notes: [a] Gum is measured in kilograms. All other products are measured in cartloads.
Source: Adapted from Jodha (1985a).

Table 3.6 Indicators of increasing scarcity of fodder and fuel in study villages in Jodhpur and Nagaur Districts

	District, by year					
	Jodhpur			Nagaur		
Details of items	1951/53	1963/66	1982/84	1951/53	1963/66	1982/84
Pachasa[a] units of fodder storage (number)	21	13	0	35	18	3
Cartloads of fodder contributed to common/charity feeding (number)	15	11	3	20	14	6
Households selling fodder (per cent):						
on exchange	28	22	8	44	28	14
for cash to trader	6	30	43	16	36	62
Duration of seasonal outmigration of sheep herders (days per year)	28	42	98	20	60	112
Time taken to collect a cart full of fuel for sale from village common (days)	10	15	40	7	12	30
Households stocking traditionally discarded crop by-products for fodder and fuel (per cent)[b]	3	27	95	6	12	98
Households allowing free access to others for post-harvest grazing, etc, (per cent)	100	56	8	100	60	15
Households replacing bush fencing (per cent):						
every year	85	52	14	88	44	16
with a gap of 2-3 years	15	47	66	12	36	59
replaced by stone fencing	–	1	20	–	10	25

Notes: There were 62 and 75 households sampled, respectively, from two villages each in the districts of Jodhpur and Nagaur.
[a] *Pachasa* is a form of staking fodder, which can be stored for 5–8 years. Literally, the word refers to the means of storing fodder unspoiled for 50 (*pachas*) years.
[b] Includes *bajra* husks, sesamum pods and stalks, mustard linseed stalks, and bengal gram stalk.
Source: Data collected during field studies, reported in Jodha (1968, 1986).

Decline in indigenous agroforestry

Although not subject to the same pressures as CPRs, some components of indigenous agroforestry are showing a rapid decline. The reduced seasonal access to woody perennials on private land has diminished the role of indigenous agroforestry systems and practices in ensuring production and stability of biomass in Rajasthan villages. The status of trees and shrubs in the agroforestry systems has been further diminished by overextraction and by certain technological changes in agricultural practices.

Overextraction   Traditionally, farmers maintained trees partly as a form of insurance, for use in times of severe droughts, prolonged sickness, and other periods of critical scarcity (Chambers *et al.* 1989). However, such farmer strategies for risk management have unintentionally been weakened by public relief measures (Jodha 1991). With improved access to urban markets, trees have increasingly been cut for sale as fuelwood and timber. Following large-scale felling, rather than lopping, of trees during recurrent severe droughts of the mid-1960s, and reduced protection and regrowth of trees in subsequent years, the number of trees declined significantly in several villages.

Technological changes in agricultural practices   Reduction in the size of their landholdings has compelled many farmers to concentrate on crop production, at the expense of woody perennials that occupied scarce space in the fields. However, the single most important change in farming practices that has adversely affected the status of intercropped woody perennials has been the large-scale adoption of the use of tractors in ploughing the fields. Their popularity stems from the ease and timeliness of operations they offer in an area with an effective wet period of only 15–20 days (Jodha 1974). At present, as much as 80–85 per cent of the net cropped area in different villages is ploughed by tractors. This has had two consequences. First, trees were widely removed from fields in order to facilitate the operation of tractors. Secondly, the ability of the annually harvested *ber* bushes to regrow was severely reduced due to the damage to their shallow root systems during the deep ploughing carried out by tractor cultivators. Within 4–5 years the fields frequently ploughed by tractor were left with no roots of this shrub.

Other changes affecting biomass management strategies and practices

The changes outlined above imply a reduction in the extent to which farmers are able to harness the complementarities and balances within the system that stem from the presence of annuals and perennials, and of extensive and intensive land uses.

Decline of fallowing Another factor that has obstructed the use of these complementarities and balances is the decline that has occurred in the extent and frequency of fallowing throughout the dry regions of India (Jodha and Singh 1990). This reflects both increased access to chemical fertilizers, public relief, and other external inputs, so that it has become less necessary for farmers to ensure regeneration of local resources, and the increasing pressures on the regenerative process.

Erosion of collective sharing The modernization and commercialization processes taking place in rural India today have also severely eroded the collective sharing that formed such an important component of traditional arid land strategies in which trees and shrubs featured prominently. Resource management has increasingly been taken over by new state-sponsored cooperatives. However, these are a creation of policy makers whose perceptions are very different from that of desert farmers, and are not substitutes for traditional forms of rural cooperation. Though they perform a function in channelling food, fodder and fuel to villages during periods of scarcity, cooperatives do not support the crucial place of biomass management and regeneration in people's overall strategies.

Less storage and recycling of biomass The reduced diversity of farming systems, and penetration of market forces that do not give a high premium to low-value labour-intensive recycling practices, have meant that options emphasizing storage and recycling of biomass products are on the decline (Jodha 1985a,b). Supply-led flexibility in the use and management of biomass has consequently been reduced.

Programmes to upgrade biomass resources and production

Though many public interventions have had the effect of leading to a decline in farmers' biomass-centred strategies, and in the role of trees, several public programmes initiated in the arid region of Rajasthan have been directed towards growth and production stability of biomass. These include programmes relating to rangelands, village woodlots, and wastelands.

The key problem with these programmes is that they are narrowly focused on particular sectors. This is in sharp contrast to the traditional integrated strategies of farmers, with their exploitation of complementarities within annual–perennial plant-based systems and balances between extensive and intensive uses of land (Chambers *et al.* 1989; Jodha 1990). Each implementing agency tends to focus on a particular product without consideration of how this fits into the overall supply of biomass and its seasonal spread. Thus, community forestry projects give scant attention to grasses, and village pasture development programmes are likely to ignore trees and shrubs.

Another related feature of the productivity-raising initiatives for CPRs is their almost exclusive focus on production technology (Shankarnarayan and Kalla 1985; Gupta 1987; Jodha 1990). Based as they are on a strong input from the relevant science or technology, these programmes emphasize techniques rather than community involvement and the user-perspective. Hence, one comes across long inventories of technically well-accessed species of trees and grasses, methods for reseeding rangelands and reforesting wastelands, plant establishment and thinning techniques, and a variety of other silvicultural or agrostological recommendations for community lands. However, there is little that explores the relevance of these measures to the institutional capabilities and sensitivities of the situation within which they are to be employed. Moreover, in order to establish and demonstrate the viability of technological measures, community lands have often been transferred into the control of the agency implementing the pilot project, thus alienating them from the people (Chambers *et al.* 1989).

3.4 Consequences and adjustments to change

The basic changes that have occurred are thus a decline in CPRs, and reduction in crop–fallow rotation and indigenous agroforestry practices. These changes have in turn adversely affected collective security of biomass, supply-led flexibility in biomass management and crop–livestock based mixed farming. The key consequence of the above changes is increased scarcity of biomass.

In response, there has been some restructuring of people's strategies in order to adjust to the new situation. These adjustments are of two types. The first relates to changes in the patterns of demand for and usage of biomass as scarcities deepen. The second takes the form of revised or new approaches to biomass production and management.

Adjusting to increasing scarcities

The data at three points of time presented in Table 3.6, from the village and farm level studies in Jodhpur and Nagaur districts, indicate the types of change that have been taking place as supply of biomass has declined over the last three decades. The practice of fodder stocking and villagers' contributions to communal and charity feeding have declined drastically, while the duration of seasonal outmigration of sheep herders and the time taken for collecting a cartload of fuel, have increased.

Table 3.6 also identifies some shifts that could be considered as indicators both of decline in the supply of biomass as well as adjustments to the decline. Thus, a shift towards increased use of cow dung in place of wood as fuel was also observed in the study villages. Another indicator of people's increasing acceptance of inferior options is that a number of biomass items like *bajra* husks (used as fodder) and sesamum stalks (fuel), which were traditionally discarded and allowed to rot, are now stocked by the people for use. Even rich farmers do not discard these products now.

The trend towards privatization of products of seasonal CPRs on crop fields is another form of adjustment to increasing scarcity. The decline in the extent and frequency with which old farm fencing is replaced by material from *khejri* trees or *ber* bushes also reflects adjustment to reduced supplies of the biomass.

One of the most important forms of adjustment to scarcity of biomass can be seen in the significant changes in livestock composition. In the study villages the average size of animal holding has declined, and the proportion of unproductive animals has been reduced. There has been an increase in the share of sheep and goats, which can manage on degraded pastures better than cattle, and can be more readily moved seasonally to canal areas of neighbouring states of Punjab and Haryana. The proportion of buffalo has also increased; a change that can be attributed to improved milk marketing facilities and better provision of drinking water in the villages since the early 1970s (Jodha 1985a). Such milch animals seldom graze on CPRs, and in a way they represent a withdrawal of rich farmers from use of CPRs.

Shifting from common to private sources of biomass

One important way in which people's strategies are restructured in the face of emerging biomass scarcities relates to changing approaches to CPR use. The widespread practices of reducing access to seasonal CPRs, and of intensifying harvesting from remaining CPRs, have already been mentioned. An important additional change is towards deliberately reducing dependence on CPRs by substitution of production from private property resources (PPRs) for the supplies traditionally obtained from CPRs.

Data from the study villages show that over a 20-year period there was a marked shift from CPRs to PPRs for collection of fodder, fuel, and food items such as gum and wild fruits. The proportion of households with exclusive dependence on CPRs for such products came down from 50 per cent or more in 1963–65 to 15–20 per cent during 1982–84. Combined use of CPRs and PPRs became more important, and more than a third of the households (compared with 3–5 per cent in the past) now depend exclusively on PPRs for collection of products conventionally supplied by CPRs (Jodha 1968, 1986).

Though there has been no revival in management of CPRs comparable to the revival in interest in agroforestry, in some areas formation of user groups through the help of non-governmental organizations (NGOs) and well focused area development programmes have resulted in some improvement. Experience with areas of privatized CPRs that have been maintained under natural vegetation have shown that modest investment in trenching, soil working or fencing, and regulated use through rotational grazing or cut-and-carry rather than open grazing, have resulted in rapid increases in productivity (Jodha

1990). The constraint to improving the productivity of CPRs through such measures is the lack of incentive to do so resulting from their open access status.

Revival of agroforestry

Mention has already been made of the overexploitation of woody perennials on fields for sale as fuel and timber, and their removal or destruction as tractor operations spread. However, a succession of low rainfall years, and the decline in availability of tree products from common lands, has increased appreciation of the importance of trees and shrubs.

The revival in interest has focused on the *khejri* tree. This is largely because, given the advantages of timeliness of field operation of using tractor, and the progressive decline in the numbers of draught animals in the cattle population as the costs of maintaining them increase (Jodha 1990), so that dependence on tractors is rising, any increase in the stock of woody perennials in crop fields must be adapted to the needs of tractor operation. This excludes the revival of the *ber* bush population. Because of its shallow roots, its occurrence as scattered colonies, and the absence of readily visible above-ground growth at the time of ploughing, it is very difficult to protect.

Managing *khejri* as an intercrop on the other hand can be feasible, as it is deep rooted, and occurs as single-stemmed trees that can be selected to form lines or other groupings that present minimum inconvenience to tractor operations. The experience of eastern districts of the arid zone (namely Sikar, Jhunjhunu) had already demonstrated the importance of *khejri* in the crop fields. Consequently, farmers in districts like Nagaur and Jodhpur started protecting and pruning (but not planting) *khejri* trees. Table 3.2 illustrates from the study village data in these two districts how, over a 20-year interval, the density of *khejri* trees increased substantially, while that of *ber* fell sharply. A more detailed study of 20 selected plots in one village showed that the increase in the number of *khejri* trees was due to a very large increase in the number of young trees less than 5 years old (Table 3.7). The impact of the changes is reflected in the shares of fodder production coming from *khejri* and *ber*.

3.5 Discussion

The fragility of the arid region's natural resources base, and its susceptibility to rapid erosion due to intensive use, to the point of desertification, has led to study after study recommending better management with greater vegetative cover (Rai 1942; Anon 1960; National Council of Applied Economic Research (NCAER) 1965; Jodha and Vyas 1969; Indian Council for Agricultural Research (ICAR) 1977; Spooner and Mann 1982). Specific suggestions

Table 3.7 Changes in stocking of *khejri* trees and *ber* bushes on 20 selected plots in a village of Nagaur District

Woody perennial components of agroforestry	Situation in monitored plots, by year	
	1963/65	1982/84
Number of *ber* bush colonies in plots[a]	168	36
Number of *khejri* trees in plots[b]		
older than 10 years	85	112
less than 5 years old	27	82
Fodder production[c] (hundreds of kg/ha) from;		
ber bushes	15	2
khejri trees	3	7

Notes: [a] Contiguous spread of *ber* bush shrub covering minimum area of 33.4 m² (40 square yards) in the same plot.
[b] Ages of *khejri* trees were distinguished by examining particular aspects of the tree such as its size, the softness of its leaves and the density of its thorns, and the size of scars on its main branches caused by annual lopping. Young trees are pruned but are seldom lopped for fodder.
[c] Fodder and fuel data were provided by plot owners and are not based on actual measurements.
Source: Data collected by physical measurement, monitoring, and mapping of 20 plots, ranging in size from 0.5 to 3 ha, reported in Jodha (1968, 1974, 1986).

have ranged from formation of a state level Board of Land Management to restrictions on crop cultivation on sub-marginal lands or controls on growth of animal numbers.

Despite these efforts, the situation of vegetative cover, and of biomass production and usage, has only worsened with the passage of time. The ineffectiveness of the initiatives can be attributed to their overemphasis on 'techniques' (i.e. mechanical or biological dimensions), and their insensitivity to institutional factors (Jodha 1988, 1990). These programmes betray complete lack of understanding of the factors and processes at village and farm level that are responsible for rapid loss of vegetation or biomass in the arid region.

Realistic ways to promote sustained availability of biomass in the arid areas of western Rajasthan need to reflect the traditional or ongoing approaches of the farmer to biomass issues. This is not to idealize the tradition, but to use it as a source of potential options, which could be improved with the help of modern advances in the field of technology and management. The paragraphs that follow summarize the key factors influencing farmer decisions about the maintenance and growing of trees on their land that emerge from the preceding assessment.[5]

The main factor driving the shift to greater reliance on private management of trees has been the decline in the extent, productivity and management of tree stocks on CPRs. There seems little scope for improvement in this situation unless State policies towards privatization change, public interventions become more sensitive than the present technical measures, and more favourable and effective user group institutional arrangements are able to emerge at the village level, with the help of NGOs.

Pressures to increase tree stocks on private lands could diminish if demand for tree products slackened. In areas with improved milk marketing facilities, reduction in the numbers of animals, changes in livestock composition, and a trend towards more intensive livestock management with stall feeding, are reducing demand for grazing. However, this is most prevalent among larger farmers, able to depend on private pasture and fodder resources. The tradition of dependence on free supplies from CPRs, and of security through large numbers of low productivity animals, are constraints to such shifts among the poor.

On the supply side, the production potential of stocks of woody plants in existing natural vegetation on farmland could be exploited more fully, and the productivity of private land kept under natural vegetation could be increased through modest investments in manuring and soil conservation measures. At present, pursuit of these options is constrained by shortages of labour during good rainfall years, and by difficulties in protecting and capturing the gains for oneself because of the convention of common access to private fallow lands. This situation might change if suitable high productivity silviopastoral systems could be introduced, or if management of trees and shrubs becomes highly profitable.

In contrast, reasonably high chances exist for promoting trees in fields, as a response to a felt need for drought period insurance, especially after the presence of *ber* bush stocks in fields declined following tractorization. Quite encouraging signs are revealed by increases in the number of *khejri* trees in crop fields. Measures to encourage this could include discouragement of tractor cultivation, and agroforestry programmes that are compatible with farmers' approaches to biomass production.

In summary, land use in the arid region of western Rajasthan is characterized by depletion of CPRs resulting from erosion of social sanctions against cutting of trees, and conversion of CPRs into open access resources where individuals overexploit the resources with no concern for their upkeep. The upsurge in tree growing on farm land is a response both to this loss of tree stocks on CPRs, and to the potential for private reward for individual effort on private property.

Notes

1. The paper is based on work carried out over a period of two decades, during which the author worked for the Central Arid Zone Research Institute

(CAZRI), Jodhpur, the Agro-Economic Research Centre (Vallabh Vidyanagar), and the International Crops Research Institute for the Semi Arid Tropics (ICRISAT), Hyderabad. Some of the results have already been reported on elsewhere (Jodha 1968, 1974, 1978, 1985a,b, 1986, 1988, 1990). The methodological details and central goals of the field work carried out during the early 1960s, early 1970s, and early 1980s, are reported in Jodha (1988). These investigations focused on wider issues of management and development of arid lands in India rather than on specific issues of tree management. Hence, in this paper the latter is discussed as an integral part of people's biomass-centred resource management strategies.

2. Using *tehsil* (subdivision of a district) level information on soil, vegetation, and water resources, a team consisting of soil scientists, agronomists and conservation specialists, has divided the region into three zones on the basis of the extent of lands of different use capability classes (Anon 1960). The bulk of the region (Zone I) consists of the lands belonging to land classes VI and VII, characterized by extremely poor soils, low rainfall, very short growing season, and high frequency of rain failures. These lands are suitable only for pasture and range development according to the Food and Agriculture Organization of the United Nations use capability classification of lands for conservation purposes. The quarter of the area of the region in Zone II is only slightly better, containing different proportions of lands of classes VI and VII and some land of classes IV. The last could be put to restricted crop cultivation using conservation measures and crop–fallow rotation practices. Only the remaining area in Zone III contains lands suitable for normal crop cultivation.

3. The term 'folk agronomy' refers to traditional or indigenous natural resource management and production practices that have been evolved and inherited by rural people without inputs from technologies derived from formal agricultural research. 'Ethno-engineering' similarly refers to indigenous measures of a structural rather than biological nature, such as the construction of contour bunds or water harvesting structures.

4. This was for all sample farmers and not only the rural poor as reported earlier (Jodha 1986).

5. This is based on a detailed submission made to the Planning Commission of the Government of India (Jodha 1988).

References

Anon (1960). *Report of the State Land Utilization Committee, Government of Rajasthan*. Government Press, Jaipur.

Chambers, R., Saxena, N.C., and Shah, T. (1989). *To the hands of the poor: trees and water*. Oxford and IBH Publishing, New Delhi.

Gupta, A.K. (1987). Why poor people do not cooperate? A study of traditional forms of cooperation with implications for modern organisations. In *Politics and practices of social research* (ed. G.C. Wanger). George Allen and Unwin, London.

ICAR (1977). *Desertification and its control*. Volume presented to the UN Conference on Desertification, Nairobi, Kenya.

Jodha, N.S. (1968). Capital formation in arid agriculture: a study of conservation measures applied to arid agriculture. PhD thesis. University of Jodhpur.

Jodha, N.S. (1974). A case of the process of tractorisation. *Economic and Political Weekly, Review of Agriculture*, **9**, A. 111–18.

Jodha, N.S. (1978). Effectiveness of farmers' adjustment to risk. *Economic and Political Weekly, Review of Agriculture*, 13, A.38–48.

Jodha, N.S. (1985a). Population growth and the decline of common property resources in Rajasthan, India. *Population and Development Review*, 11, 247–64.

Jodha, N.S. (1985b). Market forces and erosion of common property resources. In *Agricultural markets in the semi-arid tropics*, Proceedings of an International Workshop, pp. 263–77. ICRISAT (International Crops Research Institute for the Semi-Arid Tropics), Hyderabad.

Jodha, N.S. (1986). Common property resources and rural poor in dry regions of India. *Economic and Political Weekly*, 21, 1169–81.

Jodha, N.S. (1988). *Fodder and fuel management systems in the arid region of western Rajasthan*, Report submitted to the Study Group on Fuel and Fodder. Planning Commission, Government of India, New Delhi.

Jodha, N.S. (1990). Rural common property resources: contributions and crisis. Foundation Day Lecture, Society for Promotion of Wasteland Development, New Delhi. *Economic and Political Weekly, Quarterly Review of Agriculture*, 25, A. 65–78.

Jodha, N.S. (1991). Drought management: farmers' strategies and their policy implications. *Economic and Political Weekly, Quarterly Review of Agriculture*, 26, A. 98–104.

Jodha, N.S. and Singh, R.P. (1990). Crop rotation in traditional farming systems in selected areas of India. *Economic and Political Weekly, Quarterly Review of Agriculture*, 25, A. 28–35.

Jodha, N.S. and Vyas, V.S. (1969). *Conditions of stability and growth in arid agriculture*. Agro-economic Research Centre, Vallabh Vidynagar, Gujarat.

Mann, H.S. and Saxena, S.K. (ed.) (1980). *Khejri*, Monograph No. 11. Central Arid Zone Research Institute (CAZRI), Jodhpur.

Michie, B.H. (1986). Indigenous technology and farming systems research: agro-forestry in the Indian desert. In *Social sciences and farming systems research: methodological perspectives on agricultural development* (ed. J.R. Jones and B.J. Wallace), pp. 221–44. Westview Press, Boulder.

NCAER (1965). *Agriculture and livestock in Rajasthan*. NCAER, New Delhi.

Rai, R. (1942). *Akal Kashta Niwarak*, a report on famine-scarcity eradication to the counsellor of Jodhpur State (in Hindi). Bali, Marwar.

Shankarnarayan, K.A. and Kalla, J.C. (1985). *Management systems for natural vegetation*. Central Arid Zone Research Institute (CAZRI), Jodhpur.

Spooner, B. and Mann, H.S. (1982). *Desertification and development: dryland ecology in social perspective*. Academic Press, London.

4 The shift of tree cover from forests to farms in Pakistan: a long and broad view

Michael R. Dove

4.1 Introduction

Development of on-farm forestry in the Third World has been characterized, and handicapped, by a circumscription of vision. Development plans have tended to focus on the material world, ignoring the fact that ideas—the human conception of this world—are an important determinant of the dimensions of the tree niche. Development planning also has tended to focus on the village, neglecting extra-village factors that promote or circumscribe tree cultivation. Finally, development planning has tended to focus on the present, forgetting that we must understand what environmental relations were in order to understand what they are, and forgetting also that these relations are not static: they were different in the past and will be different again in the future (cf. Castro 1991; Tucker 1984, p. 341). The development of farm forestry has suffered the common curse of the immediate, local, and material development horizon—ignoring removed, distant, and non-material determinants.

The relationship between farmer and tree is comprehensible only if the scope of analysis is extended beyond the farmer and the tree. The analysis must extend to history: the rural communities that are the subject of forestry development efforts are not 'timeless' communities; they have a past that has determined their present and also will affect their future. The analysis also must extend to political economy: farmer relations with broader political and economic structures, and subjective State interest in forestry development, invariably affect farmer participation in tree cultivation. Finally, the analysis must encompass ideology: farmer participation in tree cultivation typically is perceived differently by farmers and State, and this difference can itself affect the nature of participation. In short, the involvement of farmers in tree cultivation can be comprehended only from a perspective that extends *beyond* the time and space of the village.

This chapter begins with a sketch of Pakistan's current forest cover and the historical, anthropogenic processes that have given rise to it. In the next section contemporary on-farm tree cultivation is discussed, and farmers' perceptions of the largely material needs and constraints that drive it. In the third section government efforts to support farm forestry development are analysed, together with the problems that have beset these efforts and the source of

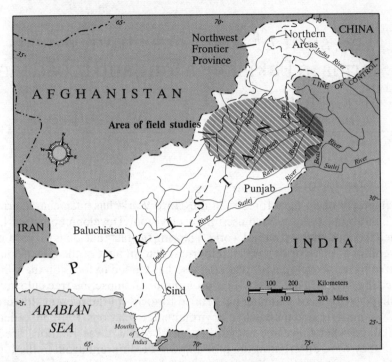

Fig. 4.1 The study area.

these problems in a self-interested dichotomization of forest and farm. The chapter concludes with comments on the broader dimensions, especially the political and economic dimensions, of farm forestry development.

The analysis is based on data gathered in stages during 1986–89, under the auspices of the Forestry Planning and Development Project, in selected districts of Pakistan's Baluchistan, North-West Frontier, and Punjab provinces (see Fig. 4.1), with a sample ranging up to 1132 households in 118 villages.[1] Data were gathered on on-farm cultivation of trees, farmers' reactions to a proposed programme to support this cultivation, the production and consumption of fuel, fodder, and timber, and farmers' perception and classification of the rural environment. Data also were gathered on the perceptions and beliefs of government officials regarding these matters (see also Dove 1992a).

4.2 The history of society–forest relations in Pakistan

Pakistan's tree cover

The climatic climax vegetation of most of Pakistan is tropical thorn forest, 'an open low forest in which thorny usually hard-wooded species predominate', which merges into dry subtropical evergreen forests in the hilly regions to the north and western part of the country (Champion *et al.* 1965, p. 111). The robustness of this vegetative cover is reflected in the varied animal life that it formerly sustained. As recently as the seventeenth century, Moghul court records and European travellers documented the presence of Asian elephant, rhinoceros and lion on the Punjab plains (Bernier 1891, pp. 374–82; Rao 1957, pp. 269–70). Today these plains are nearly devoid of natural vegetation, and all three of the aforementioned animals have vanished from this part of the subcontinent. The contemporary vegetation in most of Pakistan's arid lowlands ranges from a 'scrub preclimax' at best (Champion *et al.* 1965, p. 40) to rocky wastes at the worst. The forest cover of the country has so dwindled that it affords Pakistan what is today one of the lowest annual rates of deforestation in the world (less than one-tenth of 1 per cent per annum of the forested area), by virtue of the fact that there is so little left to deforest (Repetto 1988, pp. 6–7).

The retreat of Pakistan's forest cover does not mean that there are no more trees, however. Trees do not occur in the patterns or places familiar to orthodox foresters, but they are present nonetheless. Trees abound in all graveyards and religious shrines, where they provide shade for the pious and eternal blessings for the planter. Proscriptions against tree-felling in such places are strictly followed (cf. Gold and Gujar 1989).[2] Trees are found within the enclosed courtyards of every rural home, for which they provide shade, fodder, and fruit. The greatest numbers of trees are found on the farmlands themselves: in clusters around waterholes and tanks, where Moghul rulers decreed the planting of banyan (*Ficus bengalensis*); around wells and Persian wheels, where they shade the circling oxen that provide the motive force; and in hedgerows along field boundaries, where they provide protection against the wind and against livestock incursions, yield fuel and fodder, and offer the delicious satisfaction of legal theft of nutrients from the land of one's neighbour.

The historical background

The transition from Pakistan's natural forest cover to the domesticated tree cover just described began millennia ago, at human hands. Vedic texts from the first and second millennium BC counsel the Aryan kings to occupy the *jangala* 'open, dry lands', meaning open bush or savanna (Buhler 1964, p. 227). Savanna does not occur in the subcontinent except as a result of human activity (Champion *et al.* 1965, pp. 27–8 and pp. 38–40; Whyte 1968, pp. 167, 173, 174, 188).[3] The Aryan settlers of the arid plains of what is now Pakistan were semi-pastoralists; and in the course of pursuing their subsistence system, and in order to provide a vegetative cover better suited to this system, they altered the natural forest cover of the land from thorn forest (and dry deciduous forest) to open savanna or *jangala* (see Fig. 4.2). The *jangala* was the product of their clearing the forest and subsequently maintaining an unstable vegetative succession in its stead (see also Dove 1992b).

Almost as soon as this transformation of the natural forests began, the creation of 'un-natural' forests also began (cf. Shiva 1989, p. 59). The Vedic texts contain injunctions to plant and protect not just trees or gardens, but also forests. For example, the *Kautiliya* of Arthasastra suggests that forests should be planted whenever a new state is established (Kangle 1988, p. 174; cf. Shiva 1989, p. 59). This tradition of State forestry continued (in varying degrees and fashions) up to the creation by the colonial British administration of the Indian Forest Service in the mid-nineteenth century, which has succeeded (with a considerable amount of continuity) to the Forest Service of contemporary Pakistan. There has been a folk counterpart to this tradition, especially in the frontier tribal areas where State control has been weakest, consisting of the management of village forests (variously called *shamilat* or *hazara* forests in Pakistan today).

Pakistan's State forests (and its village forests too) have largely gone the way of its natural forests, and the locus of management of tree resources is in the process of shifting again, this time to private farmlands. The loss in forest area is reflected in the decline of Pakistan's officially recognized 'forest cover' to approximately 4 per cent of the nation's land area. The increase in on-farm tree cover is not reflected in these statistics, which, as the term 'forest cover' implies, are based on traditional measures of the amount of land under blocks of forest, not the total amount of land under trees.[4] Indirect evidence of the extent of the on-farm tree cover is provided by evidence that farms provide 90 per cent of Pakistan's fuelwood needs (Leach 1987, p. 42), and by the fact that despite the near-total loss of its public forests, fuelwood prices in

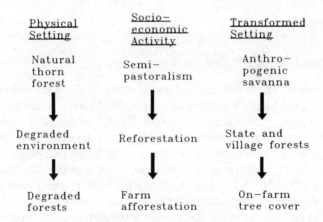

Physical Setting	Socio-economic Activity	Transformed Setting
Natural thorn forest	Semi-pastoralism	Anthro-pogenic savanna
↓	↓	↓
Degraded environment	Reforestation	State and village forests
↓	↓	↓
Degraded forests	Farm afforestation	On-farm tree cover

Fig. 4.2 Historical coevolution of society and environment in Pakistan.

Table 4.1 Sources of existing on-farm trees

Sources of trees	Per cent of households reporting
All natural	50
Natural and planted	27
All planted	22

Pakistan have remained basically stable over the past two decades (Leach 1987, pp. 54–6; Campbell 1992, p. 309). Direct evidence of the extent of on-farm tree cultivation is provided by the data gathered for this study, which show almost one-half of all farmers reporting that some or all of the trees on their farms were planted (Table 4.1).[5]

This cultivation is a recent development, as reflected in the relatively recent dates that farmers give for the commencement of tree planting in their villages (Table 4.2).[6] The recency of the forest-farm shift in tree cover also is reflected in the fact that there is as yet no distinctive term for trees planted on farms. In order to distinguish timber, fuel and fodder species, all of which are new to cultivation, from fruit trees, which have long been planted on farms, farmers call the non-fruit trees *junglot ke drukht* 'forest trees'.[7]

Table 4.2 Perceptions of the time depth of on-farm forestry

Farmers response to the question, 'When did tree planting/cultivation begin?'	Per cent of households reporting
1. Not relevant. No tree planting done.	13
2. A long time ago. The time of the ancestors	15
3. Do not know.	20
4. *X* number of years ago[a]	53

Notes: [a] The mean value of X was 25.4 years ($\sigma = 24.5$).

Conceptions of society–forest relations

The changing relationship between society and forests in Pakistan has an ideological as well as material aspect: equally important to what society did to Pakistan's forests is what it thinks it did. For example, there is no recognition in the Vedic literature of the fact that Aryan settlement and subsistence were responsible for the transformation of the natural Indian forest to *jangala* 'savanna'. Such statements as, 'Let him [the Aryan king] settle in a country which is open and has a dry climate [*gangalah* in the original]' (Buhler 1964, p. 227), imply that the *jangala* was thought to be prior to human settlement and land use. The ancient Vedic literature emphasizes the distinction between *jangala* and forest as one between the environment of civilization and the environment of barbarism, not as one between natural and anthropogenic landscapes (Zimmermann 1987). This literature presents the *jangala* as land that is chosen because it is good, not land that has become (or been made) good because it was chosen. The State's conception of its environment as ritually pure mitigated against recognition of *jangala's* anthropogenic character, thus illustrating how perception of environmental relations by the State can be 'deflected' by self-interest.

Self-interested misconception of environmental relations also characterized the colonial state. Colonial society attributed the spontaneous reforestation of agricultural areas, e.g. to inimical cultural forces. Colonial observers claimed that displacement of Hindu populations by Muslim ones inevitably resulted in the transformation of intensively cultivated 'good' landscapes into overgrown, sparsely cultivated *jungli* 'bad' ones (cf. Heyne quoted in Bartlett 1956, pp. 280–82). (Note that the nature of the misconception is not the same in colonial and ancient times: while the Aryan State associated the *jangala* landscape with the presence of civilization, the colonial government associated the *jangli* landscape with its absence, in the sense of the absence of 'rational' land use.) This was attributed to the Islamic system of agrarian taxation, which was said to penalize any long-term investment in soil fertility, because its limitations on the length of tenure discouraged long-term planning and the heaviness of its exactions encouraged periodic flight (e.g. Moreland 1988, pp. 205, 207).

The problem with this assessment is not whether the Islamic system did indeed have such an impact on land-use patterns (which may well have been the case) but whether this impact made sense within the context of a broader association of social system and land use, which also seems to have been the case. The *jungli* landscape was not the product of a flawed socio-political system, as colonial society supposed, rather it was the manifestation of a rational land use system, characteristically associated not just with a particular culture but also a particular social and physical environment. Through much of India's history there has been a fundamental dichotomy between predominantly Hindu intensive agriculturalists living in comparatively dense populations along rivers and in their flood plains (the *bet*), and predominantly Muslim extensive agriculturalists and pastoralists living in the sparsely populated arid interior regions (the *doab*). Whenever a population pursuing one of these systems displaced a population pursuing the other, some intensification or extensification of land use is likely to have followed. In any case, the colonial government's critical assessment of any movement toward extensification was clearly self-interested, since intensive land-use systems are inherently easier for central governments to administer and exploit than extensive ones.[8]

4.3 On-farm tree cultivation

The product of these historical processes is the contemporary pattern of on-farm tree cultivation.

On-farm forestry

On-farm tree cultivation in contemporary Pakistan takes a variety of forms. It ranges from not destroying naturally growing tree seedlings, to protecting them (cf. Michie 1986, p. 240), to planting them. A common form is the complex interweaving of assorted trees and grasses in the ubiquitous (but unstudied) brush fences of Pakistan. Another example is the cultivation of blocks of nitrogen-fixing trees to restore agriculturally taxed soil, with side benefits of fuelwood and fodder, in the *kikar* (*Acacia nilotica*)-based *hurries* of Pakistan's Sind Province (see Sheikh 1986). In some areas, there is actual mixed cropping of trees and food crops.[9] According to our surveys, trees most often either are scattered (singly or in small numbers) over the surface of the farm or else clustered within the walled courtyard of the house; linear and block plantings are rare (Table 4.3). The term 'linear planting' is used here in the way that it is used by Pakistani farmers, namely to refer to the single-species planting of trees along a field bund. Farmers usually do not use the term 'linear planting' to refer to brush fences, which are far more common than the 10 per cent indicated for 'linear plantings' in Table 4.3. The species preferred for on-farm cultivation are the major multi-purpose trees of the western part of the subcontinent, in particular *Acacia nilotica* and *Dalbergia sissoo* (Table 4.4). The major reasons for on-farm tree cultivation are meeting household needs for fuel, timber, and fodder (Table 4.5).

Table 4.3 Patterns of on-farm tree cultivation

Tree cultivation configurations	Per cent of households reporting
Scattered planting	74
Planting in courtyards	60
Linear planting	10
Block planting	3

Notes: Some households may have reported more than one planting configuration.

The character of on-farm tree cultivation is affected by a number of different physical and social variables, all of which affect the relative attraction to the farmer of using land to cultivate trees as opposed to using it for an annual crop, for grazing, or for nothing at all (see Dove, in press). One such variable

Table 4.4 Farmers' requests for particular types of tree seedlings

Species requested	Per cent of households requesting
Acacia nilotica	48
Dalbergia sissoo	46
Eucalyptus spp.	44
Populus spp.	17
Leucaena leucocephala	8
Acacia modecta	8
Others	18

Notes:
This table presents the per cent of farm households which requested particular species from the Forestry Planning and Development Project. Households could request more than one species.

is the location of the village: the closer it lies to State forests, from which tree products can be taken (albeit illicitly), the lower the interest in tree cultivation. Also important is the proximity of the village to vehicular roads and towns: the greater the proximity, the lower the interest in tree cultivation tends to be, because villagers usually are more involved in off-farm income-producing activities and less interested in the kind of investment in their land that farm forestry represents. Urban wage-labour opportunities (both in Pakistan and abroad) may have the opposite effect, however, in so far as they lead to

Table 4.5 Farmers' reasons for planting trees

Reasons for tree planting	Per cent of households reporting
Fuelwood	91
Timber	72
Sale	46
Fodder	13
Other	7
Uncertain	6

Notes: This table presents the reasons for tree planting given by households to the Forestry Planning and Development Project. Households could report more than one reason for tree planting.

a shortage of agricultural labour in the village: this increases the relative attraction of perennial tree crops *vis-à-vis* annual food crops, because trees yield higher returns to labour and require less labour overall.

Another labour-related variable that affects interest in on-farm tree cultivation is tenurial relations. In the study districts, an average of 39 per cent of all farms are partially or wholly operated by tenants or sharecroppers (Government of Pakistan (GOP) 1980, Table 1.2). Tenant-farming or sharecropping contracts typically run for 1 year and necessarily, therefore, exclude rights to a multi-year crop like trees. As a result, tenants and sharecroppers typically treat on-farm trees as (in their own words) 'weeds', uprooting them on sight. (One tenant graphically described how the mere sight of a tree growing on the land he farms gives him a pain in his stomach.)

Perhaps the most important determinant of interest in on-farm tree cultivation (the most important determinant of the relative attraction of using land for trees as opposed to some other purpose) involves the capabilities of the land. The ability to use land (like labour, as just discussed) for purposes other than tree cultivation, or the ability to use land for tree cultivation at the same time as it is used for other purposes, especially food cropping, is central to interest in on-farm tree cultivation. This is reflected in the fact that one of the most commonly reported farmer concerns regarding tree cultivation is the impact of this cultivation on annual crops (Table 4.12). Farmers worry about tree-crop competition not just with respect to sunlight, soil nutrients, and water, but also with respect to human labour. Farmers say that whenever tree planting competes for labour with cultivation of annual crops, they will sacrifice working with the trees in order to work with the crops. The annual agricultural activities that are most likely to conflict with tree planting are harvesting and sowing. Farmers endeavour to plant trees during those times in the winter or monsoon planting season when competition for their labour from these activities is diminished.

Interest in tree cultivation tends to be higher, therefore, on land that is not suited for food cropping.[10] Land may be unsuited to food cropping, and thus suited to tree cropping, due to problems of erosion, waterlogging, or salinization, but in a predominantly arid and semi-arid country like Pakistan the most common problem is lack of water. While water or irrigation[11] is central to interest in tree cultivation, however, the relationship between the two is not unequivocal. In the absence of irrigation, tree cultivation may be inhibited by farmer concern that trees will compete with food crops for scarce moisture; but in the presence of irrigation, the desire to recoup its high costs through cultivation of high-value food crops may deny any space to trees. This tension is reflected in the fact that our surveys found the greatest number of trees, as well as the greatest number of planted trees, on farms containing a *mixture* of irrigated and un-irrigated lands (Table 4.6). The relationship between irrigation and tree cultivation also is complicated by the type and intended purpose of the trees: whereas trees for fuelwood predominate on completely un-irrigated farms, trees for use as timber predominate on farms that are completely irrigated. This reflects the suitability of low-value fuelwood trees to the cost of production inputs on un-irrigated lands and that of high-value timber trees to the cost of production inputs on irrigated lands.

An oft-overlooked characteristic of the land, which also is relevant to farmer interest in tree husbandry, is its *natural* ability to generate a tree cover. The arid, steppe-like character of Pakistan's plains seems to suggest that the land long ago lost this ability, but this is not the case. An example of the latent capacity of the land to produce trees is given by Margalla Hills federal park, which adjoins the national capital of Islamabad. The erection by the federal government of huge 'hurricane' fences all around the perimeter of this park has greatly reduced (albeit not eliminated) the pressure of livestock on the local vegetation, as a result of which a robust natural vegetation, an incipient thorn forest, has sprung up. In the quarter century since Islamabad was founded and the park laid out, this vegetation has already surpassed in luxuriance almost anything that can be seen on the plains of Pakistan. This success demonstrates that the barren vegetative cover in much of contemporary Pakistan is artificial and is due to grazing by livestock of regenerating trees.

The cost of fencing and the need for grazing lands make this method of natural afforestation impractical for the vast majority of Pakistan's farmers, but not other methods, one of the most common of which is found on Pakistan's *bet* 'riverine' lands. Annual or semi-annual floods disperse the seeds of native trees (especially *Acacia nilotica*) over these lands, which then take root and, if left alone by the local human and animal population, grow sufficiently in one season to survive the succeeding floods and eventually constitute a forest. (This is the genesis of the dense riverine forests of the lower Indus, renowned to this day as the sanctuary of dacoits.) The owners of such

Table 4.6 Extent of irrigation and on-farm tree cover

| Extent of on-farm tree cover | Extent of farm irrigation | | |
	All unirrigated	Mixed irrigated/ unirrigated	All irrigated
None	20	12	7
Some	80	88	93
All natural	58	48	32
Mixed	25	27	34
All planted	17	25	34

Notes: This table presents the percentage of households in each irrigation category reporting whether there are trees on the farm or not and, if there are, whether they are planted, natural, or both.

forests usually sell them, at maturity, on a standing stock basis to itinerant Pathan contractors. After the felling has been carried out, the natural cycle of afforestation is allowed to begin again.

There are even areas on Pakistan's plains where the dynamics of natural afforestation are so robust as to pose not an opportunity for, but a threat to, the local farmers. In the Son Valley, in Punjab's Khushab District, farmlands are surrounded by vigorous scrub forests of *Prosopis juliflora* (an exotic said to have been introduced into Pakistan early this century by the Forest Service). The growth of this tree is so vigorous that the local farmers must devote considerable labour to keeping it out of their fields, sometimes failing in the attempt. Indeed, the farmers spoke of *Prosopis* as the principal constraint on their food-crop cultivation. Because the farmers of this region regard a 'tree' as their chief crop 'pest', and because of the additional unfortunate circumstance that it is locally called *jungli kikar* 'wild Acacia', the farmers showed no interest in the Forestry Planning and Development Project offer of assistance in tree cultivation, and they reacted with positive hostility to the Project's offer of *kikar* (*Acacia nilotica*) seedlings.

Farmer perceptions

As the story of the 'wild Acacia' demonstrates, farmer perceptions of their past and present relations with trees affect what they will do in their future relations. In order to understand their current and future participation in afforestation, therefore, it is important to understand what farmers think about past deforestation.

Farmers attribute deforestation equally to too little replanting (both by farmers and by the Forest Department) and too much cutting (Table 4.7). The farmers, both explicitly and implicitly, associate deforestation with resource pressure. This also is clear in Rauf's (1981) survey of farmer perceptions of the causes of changes in vegetative cover in northern Pakistan (Table 4.8). In areas where the tree cover is not decreasing but increasing (referring for the most part to private farmland as opposed to State forest land), farmers attribute the increase to more planting, made possible by market forces, increased awareness or interest and, especially, increased access to irrigation (Table 4.9). The importance of irrigation in promoting increased tree cultivation is reflected in a positive association between village irrigation and a perceived increase in tree/forest cover (Table 4.10). While farmers identify *increased* usage of tree products as a major determinant of decreases in tree/forest cover, they do not attribute increases in tree/forest cover to *decreased* usage (Table 4.11). That is, farmers do not identify diminished usage, or conservation, as a major cause of increased tree/forest cover; they attribute it solely to increased planting.

Farmers take the same view of on-farm tree cultivation that they take of the broader issues of deforestation and afforestation; they perceive resource

Table 4.7 Perceived determinants of decreases in tree cover, mostly in forests

Perceived determinants	Per cent of responses
Decreased production by farmers and/or by the Forest Department	57
Because of:	
less planting by farmers and/or the Forest Department	33
other failures of the Forest Department	9
lack of land for tree planting by farmers	7
lack of interest in tree planting by farmers	4
other	4
Increased usage of trees	43
Because of:	
more cutting	35
more population	7
selling to markets	2

Notes: This table presents the responses reported to open-ended questions about why tree cover has decreased.

pressure, in effect, to be its major constraint (Table 4.12). Rauf (1981), in contrast, found lack of need for tree products to be the major obstacle to farm forestry in the heavily forested hills of northern Pakistan (Table 4.13). More similar to our findings are the results of two surveys from Rajasthan, which show that farmers' major perceived constraints on tree cultivation are

Table 4.8 Perceived determinants of changes in ground cover, Hazara District, northern Pakistan

Determinants of change	Per cent of responses
Resource pressures	94
Due to:	
increased cultivation	44
illegal cutting	40
growing population	9
land reform	1
Other	6

Notes: Adapted from Rauf (1981, p.63).

Table 4.9 Perceived determinants of increases in tree cover, mostly on farms

Determinants of increases	Per cent of responses
Increased production by farmers and/or the Forest Department	97
Because of:	
more irrigation	31
more planting	21
more interest	21
market potential	14
decline of the commons	7
new species available	3
Decreased usage of trees	3
Because of less cutting	3

Notes: This table presents responses reported to open-ended questions about why tree cover has increased.

lack of land, labour, and water (Agarwal 1986, p. 132). The most commonly cited obstacle to tree cultivation, in our surveys, is lack of water. In some of the most arid parts of Pakistan's plains, farmers responded to the invitation to participate in the Forestry Planning and Development Project's planting programme by saying *Pina ke pani bhi nehi* 'Even drinking water is not available', much less water to give to trees. The second most oft-cited obstacle is difficulty of protection. Problems of tree protection are exacerbated by the great fragmentation of farmland in Pakistan,[12] and also by the widespread

Table 4.10 Association between irrigation and perceived increases in forest cover

Perceived village trend in forest cover	Village source of water for cultivation	
	Rain-fed	Part/whole irrigated
Decreasing	15.5	9.5
Increasing	3.5	12.5

Notes: $n = 41$ villages. $X^2 = 4.81$, significant to 5 per cent

Table 4.11 Changes in tree planting and tree-use and changes in tree and forest cover

	Per cent of responses
Perceived determinants of decrease in tree and forest cover, mostly in forests	
because of less tree planting	52
because of more tree use	48
Perceived determinants of increase in tree and forest cover, mostly on farms	
because of more tree planting	97
because of less tree use	3

practice of rotating farmland on a 2-year cycle between cropping and grazing. The third most frequently cited obstacle to tree cultivation is fear of a deleterious impact on the annual food crops, through competition for light, land, and water (Table 4.14). This problem is exacerbated by the aforementioned fragmentation of land and shortage of water.

Table 4.12 Perceived constraints to on-farm tree cultivation and management

Constraints to tree cultivation and management	Per cent of responses
Resource competition	73
a lack of water	31
protection problems	21
threats of impacts on crops	21
Other constraints	24
lack of seedlings	14
pests and diseases	7
soil problems	3
No constraints	3

Notes: The nature of this particular enquiry did not extend to other important constraints, such as institutional factors, which can limit the ability of farmers to cultivate and management trees. As Chambers and Leach (1989, p.340) observe, '. . . The major obstacle to tree planting by small and poor farmers on their lands lies not with them but with officials and other outsiders, with laws and regulations, and with their implementation.'

Table 4.13 Farmers' reasons for failure to cultivate trees, Hazara District, northern Pakistan

Limiting factor or types of contraints	Per cent of responses
Farmer	71
no need or time	35
has access to natural forests	12
lacks the knowledge	9
perceived competition with grass	9
unable to provide care	6
Government	27
fails to provide seedlings or to otherwise cooperate	18
fails to demarcate land	9
Land too hilly	2

Notes: Adapted from Rauf (1981, p.72).

One other major constraint to on-farm tree cultivation is the lack of seedlings (Table 4.12). The accepted wisdom in social forestry circles is that seedlings are never in such short supply as to pose a constraint to tree cultivation, but Pakistan's plains appear to present an exceptional case. One of two principal reasons for this involves the special significance of trees within the social ecology of tenant farms. As stated earlier, tenants have no rights to trees and treat them as hateful weeds. The second reason is the abundance of livestock,[13] especially goats which outnumber cattle two-to-one (Mohammad 1989), which vigorously suppress the natural on-farm stock of tree seedlings. As the reference to the semi-pastoral Aryans at the start of this study indicates, livestock have long been a central part of human subsistence systems in this part of the subcontinent. Even today in many parts of Pakistan, the primary goal of what appear to be food-crop-oriented farming systems is the production of fodder for livestock (Supple *et al*. 1985, pp. 33–4, 51–3, 61–2). In such systems, the niche for perennial tree crops may be highly restricted. One of the chief reasons for this is the tendency to practice free-grazing versus stall-feeding, which has roots in Pakistan's pastoral past and represents an adaptation to the relative poverty of its natural environment.

The cropping system called *do fesela, do sol* 'two crops, two years' (cf. Supple *et al*. 1985, pp. 31, 41) illustrates how free-grazing restricts the on-farm tree niche. In this system the lands of an entire village are divided in two; for one-year, one half is used for crops and the other half for free-grazing, and the next year the uses are reversed. This 2-year rotation provides tree seedlings only a 1-year 'window-of-opportunity' in which to establish themselves and

Table 4.14 Perceived impact of tree cultivation on food/crop cultivation

Objects of competition between trees and crops	Per cent of responses
Sunlight	39
Land or space	31
Water or soil moisture	30

Notes: This table presents responses reported to an open-ended question about the perceived impacts of planted trees on crops.

thus is highly inimical to tree growth. The scarcity of seedlings is recognized by the government in Pakistan, as reflected in the annual seedling distribution that it sponsors (as the principal act of public extension in its reforestation campaign).

4.4 Policy implications

Interventions: real and ideal

The government's annual seedling-give-away has enjoyed only limited success, and the reasons for this are illustrative of broader problems with forestry policy. The programme usually does not target the population most in need of its assistance: the poorer, less accessible farmers and landless tenants. Also, the programme's emphasis, with the farm population that it reaches, is on the delivery of tree seedlings and a motivational message. The implication is that the principal obstacle to farmer involvement in tree cultivation is lack of planting stock and absence of motivation. In fact, farmers are undertaking substantial on-farm afforestation on their own, and the problems with which they need help have less to do with psychological motivation and more to do with material constraints, problems that, seedlings aside, are not addressed in the annual planting campaigns. Seedlings alone are not, in any case, sufficient for afforestation. Foresters in Pakistan, as elsewhere in the world, interpret deforestation as literally a loss of trees, which they accordingly try to rectify with the provision of trees. In fact, deforestation is caused not by a loss of trees, but by the loss of a *niche* for trees within the social, physical, and political environment, which is both a very different, and less easily remedied, loss.

It might be more accurate to say that the tree niche has not been lost but rather transformed: historical trends in environment and society in Pakistan (as in many developing countries) increasingly favour the growth of trees on private farms as opposed to State forests (Dove 1993). Government forestry programmes that are congruent with this trend, this shift in tree cover from forests to farms, will tend to be more successful than those that are not. An example of the latter is the government's efforts, widely supported by international development organizations, to reforest village commons in Pakistan. Such efforts are in conflict with the village-level forces that are making tree-growing more viable on private farmland than on common or public land. It is little wonder, therefore, that these efforts usually prove problematic (cf. Cernea 1985), as reflected in the fact that their management schemes typically break down when their international sponsors withdraw their support (cf. Dani 1988, p. 298).

Another example of both the weaknesses and strengths of government policy is the Forestry Planning and Development (FP&D) Project, Pakistan's first nation-wide social forestry project. Its goal was to expand tree planting and, thereby, the production of fuelwood, fodder, and timber on farmlands in Pakistan, thus improving the long-term economic and ecological viability of small farms. Its field component was designed to provide small farmers with extension advice and free seedlings for planting on their own lands. The design of this project was very much in line with the new tree niche, but its application was less so. Many of the project's foresters, instead of contacting small farmers and supporting interest in small, subsistence-oriented plantings of indigenous multi-purpose trees, initially contacted large landlords, whom they tried to interest in large market-oriented plantations of fast-growing exotics like *Eucalyptus* spp. and *Populus* spp.[14] The intended target group of small farmers was dismissed as 'non-progressive' farmers who could not be expected to be interested in participating in farm forestry. The manifest evidence that such farmers were *already* engaged in farm forestry, foremost among which is the ubiquitous thorn fence of Pakistan,[15] was not recognized by most foresters.

Also not recognized was the small farmers' need for farm forestry. Many project foresters attributed signs of on-farm shortages of tree products, such as the use of dung for fuel as opposed to fertilizer, to cultural idiosyncracies as opposed to material need. In selecting farmers for participation in the project, thus, it was not need but its absence that tended to rule. There was conscious selection for farmers with prior experience with agroforestry (which obviated the need for extension advice) and who were well-off, which meant that they had the ability to purchase seedlings if they had not been given them by the government, and which also meant that they did not have to worry about shortages of fuelwood, fodder, or timber.

The problem

The farmers' interest in and need for tree cultivation was not recognized because, first, the patterns of this cultivation lie outside the orthodox forestry mould. (The magnitude of this difference can be seen when one tries to apply textbook methods of forest mensuration, e.g., to a thorn fence instead of a forest.) This does not explain all of the difficulty, however. The principal reason why foresters cannot recognize the farmers' actual stance toward tree cultivation is that they *must not* recognize it. While farmers have no reason to obfuscate the material conditions of their stance toward tree cultivation, foresters do: it is inconsistent with the ideology that traditionally supported their institutional reproduction. This ideology rejects the farmers' perception of material factors as the primary factor in deforestation and afforestation; it proposes in its place an explanation in terms of cultural factors.[16]

Forestry orthodoxy in Pakistan traditionally attributed illicit felling in forested areas to the 'anti-tree' attitudes of the rural people (cf. Agarwal 1986, p. 108). State foresters up until recently maintained that farmers were not merely the enemies of the forest (namely, the State's forest), but that they were also opposed to trees *per se* (even on their own farms). According to accepted wisdom, the farmers were not 'tree-minded'. These explanations all invoke a rural belief system that is unsuited to sustainable resource use, as opposed to a rural economic and political context that pressures and contests the resource base; they construe conflict between forester and farmer as conflict between cultures or world views, as opposed to competition between equally self-interested resource users. Such an explanation is self-interested: it is easier for foresters to defend State forests against tree-felling by farmers, e.g., if the tree-felling is ascribed to a 'bad attitude' rather than a lack of fuel for cooking.

This in-attention to the material basis for deforestation is reflected in the government emphasis on public motivation and education (as in the annual tree-planting campaign), as opposed to reducing the imbalance between the resource demands of the State and the public, on the one hand, and the resource supply on the other. This dichotomy between the material world of the farmers and the material world of the forests is associated with a traditional limitation of Forest Department operations to State forests. Under the first three post-Independence forestry policies of Pakistan, the Forest Service had no mandate to work outside these forests: not on farmlands, not even on the country's rangelands. The latter exception is particularly telling, since rangeland comprises approximately two-thirds of Pakistan's total land area, and it fell under the aegis of no other agency.[17]

This dichotomy between farmer or farm on the one hand and foresters and forests on the other is reflected in the interesting distinction between so-called 'forest' and 'fruity' trees. The Forest Service classifies fruit trees as 'non-trees': the forest laws state that 'the expression "tree" . . . does not include . . . any fruit tree' (GOP 1962, p. 72). This law both reflects and supports allocation to the Agriculture Department of responsibility for 'fruity trees' on private farms. The term 'fruity tree' generally is used to refer to any trees grown on farms primarily for their fruits (such as orange or mango); while the term 'forest tree' refers to all other trees (see note 5). While 'forest tree' connotes a tree whose principal products are timber, fuelwood, or fodder, many also bear fruit, in some cases fruit with economic importance (e.g., *Acacia nilotica* and *Zizyphus jujuba*). Further, while 'forest trees' usually are found in the forest, they also are found on farms (where they are still termed 'forest trees', however). Finally, while 'forest trees' most often grow naturally, they also may be cultivated, whether on farms or in forests.

The distinction between tree and fruit tree is not a simple one, therefore. It may be expressed best as the distinction between forest (*jangal*) and orchard (*bagh*), with the latter tending to be more private, managed, and fruit/food-oriented than the former. The latter has more to do with culture, thus, while the former has more to do with nature. This distinction throws new light on the policy of the Forestry Planning and Development Project (and other government forestry projects) to distribute only 'forest' or 'non-fruity' trees to farmers. It suggests that the intent of such projects is not just to grow trees on farms, but to reproduce the forest on farms. This also is suggested by the term coined for the project's focus, namely 'farm forestry'.

The Forest Department's interest in reproducing the forest on private lands, as opposed to integrating trees into farming systems, is reflected in its traditional pattern of interaction with the rural population. The Forest Service traditionally had two types of relationships with two distinct rural clienteles. From one clientele, the peasantry, the Forest Service extracted fees for approved uses of the forest (e.g., grazing cattle or gathering fuelwood) and fines (and bribes) for unapproved uses (e.g., grazing cattle and gathering fuelwood without paying the fees).[18] For the other clientele, the principal landlords in each district, the Forest Service provided 100 per cent subsidized tree-plantings on their lands, in the context of a broader pattern of reciprocal economic and political ties between the government and the rural elite. The scale of these plantings tended to be relatively large: a 1972 report from a government committee noted that the only major obstacle to tree planting by farmers was possible obstruction of the flight of aircraft (used to spray pesticides on the trees) over the farmers' fields (Bokhari 1989, p. 36). In scale and management, therefore, the private tree plantations that the Forest Department helped establish resembled the Department's own State forests; and the Department's distinction between peasants and landlords was, in effect, a distinction between farm and forest.

4.5 Conclusions

The lessons of this analysis concern the need to place the development of farm forestry in Pakistan in a wider context, incorporating dimensions of time, space, and social structure that are not usually considered relevant. For example, it is suggested that this development cannot be interpreted properly without reference to the prior history of forestry development. The current development is *not* the first change in a 'timeless tradition' that remained intact until confronted by twentieth century forces of modernization. Rather, it is the latest stage in a process of co-evolution of society and environment that has been taking place in the subcontinent for millennia. Similar and earlier developments included the transformation of the natural climax thorn forest into fire-climax savanna in the Vedic era; the so-called transformation by Muslims of agricultural land into 'wasteland' in the colonial era; and the effort made in all eras to create and protect forests (as well as destroy or transform them). It is important to see the development of farm forestry within this historical context, for several reasons. First, some aspects of the contemporary relations between society and forests (such as the way these relations are perceived by the State) can be illuminated by comparison with the historic relations. Second, the contemporary relations to some extent can be derived from the historic relations: what has gone before in environmental relations determines much of what comes after. And third, this history dissuades us from viewing contemporary change as unique: it shows change to be the rule rather than the exception.

Another dimension that this study adds to the understanding of farm forestry development is ideological. Examples have been presented from ancient, colonial and contemporary times, of the subjective character of State perception of relations between society and forests. In each case, this subjectivity sustained (and thus was sustained by) the self-interest of the State. This subjectivity must be seen as part of the context in which farm forestry develops, which is potentially as important as any aspect of the physical environment. The lesson from this perspective on farm forestry development, thus, is that any development of a resource as important as forests is likely to be perceived subjectively, because any such development is necessarily a political act.

As this last comment suggests, the third and final dimension of this study is political. Farm forestry development represents a response not just to local environmental conditions, nor just to socio-economic conditions within the village, but to a wider set of relationships that encompass the physical environment, rural society, *and* the State. The development of farm forestry at the same time that traditional forests are being degraded is not accidental: our surveys confirm that the loss of forests on local village or State lands is associated with farmers' decisions to cultivate trees on their own lands (Table 4.15) (cf. Michie 1986, p. 240). The association between deforestation and farm afforestation is not a response just to diminishing forest lands, however, but to diminishing control of these lands. Resource scarcity is a function of both absolute scarcity and, from the farmers' viewpoint, restricted access. The diminishment of Pakistan's State forests has been accompanied by tightened government control over what remains. Public access to village or *shamilat* forests has been similarly restricted by gradual privatization (Cernea 1985). The evolutionary response of farmers to this loss of control over tree resources has been to re-establish control by restocking the resource in a different context, to re-establish control in the only place where farmers can do so,

Table 4.15 Association between village forests and past tree-planting

Per cent of village households that in the past	Village has forest (per cent)	
	No	Yes
Did not plant trees	43	56
Planted trees	57	44

Notes: $n = 305$ households (in multiple villages). $X^2 = 8.55$, significant to 0.5 per cent.

on their own farms. The salience of this issue of resource control is reflected in the fact that the first questions invariably asked by farmers, regarding the Forestry Planning and Development project's planting programme, were 'To whom will the trees belong?' and 'Will the Forest Department take over the land on which the trees are planted?'

The possibility of loss of control of lands on which trees are planted can be seen in the Forest Department's practice, still ongoing, of assuming on a contractual basis the management of village or *hazara* forests in northern Pakistan (cf. Dani 1988, pp. 72–3). While this is not supposed to affect village ownership of the forests, cases have been reported in which such forests have been kept by the Forest Department (which then classifies them as 'resumed lands') on the grounds that they were really formerly government lands. The potential loss of control of on-farm trees was raised even more explicitly in the Punjab Plantation and Maintenance of Trees Act of 1974. Bokhari (1989, pp. 58–9) writes as follows of this Act:

This Act was promulgated by the Government of Punjab in order to promote planting and maintenance of trees on private lands. The occupier of the land was required to plant and maintain at least 3 trees per acre. In the event of his failure to do so, the Act provided for a penalty of one rupee per tree. Although the Act is still valid, it has not been seriously pursued by the department, because of a negative reaction shown by the people towards the penal clause. Perhaps the framers of the Act forgot to realize that people cannot be coerced to plant trees on their farmlands through punitive measures provided in the enactment.

The evidence that there already are trees on farms, and that farmers are not opposed to their cultivation, suggests a different interpretation of the Act and farmer reaction to it. It suggests that the Act was less an attempt to promote the shift from forests to farms, than an attempt to parry this with a shift of government control. (The much-discussed, purported opposition to tree cultivation among farmers is central to the rationale for this extension of government control: if farmers are *not* opposed to tree cultivation, this rationale evaporates.) In the context of this shift, the officially sanctioned 'planted and maintained' tree symbolizes a government effort to extend its control to private farmlands; and farmer opposition represents opposition not to trees but to this effort.

This example aptly sums up the thesis that has been explored in this chapter: the relations between farmer and tree involve much more than one individual and one plant in one place and time. They extend back in time and out in geographic, social, and political space; and they can only be correctly interpreted by taking these broader dimensions into account.

Notes

1. The study comprised five stages:
   - group interviews focusing on village characteristics in 118 villages;
   - interviews on basic household characteristics in 1132 households in 63 villages;
   - in-depth interviews on farm ecology and economics in 589 households in 40 villages;
   - in-depth interviews on village ecology with 40 groups of key informants and *maolvis* 'mullahs' in 40 villages; and
   - monitoring of daily activities for 18 months in 13 households of key informants.

   Field work was supported by the Forestry Planning and Development Project, jointly funded by the Government of Pakistan and the United States Agency for International Development, under the direction of the Inspector General of Forests and under contract to the Winrock International Institute for Agricultural Development. This analysis was supported by Yale University's Agrarian Studies Program, the East–West Center's Program on Environment, and the East–West Center's Program on Population (under USAID grant #DPE-3046-A-00-8050-00). The author is grateful to the following persons for assistance: for aid with field research, R. Ahmad, S. Ahmad, N. Ahmed, A. Hassan, N. Marvat, Z.I. Marwat, U.F. Marwat, Z. Masood, S. Qamar, J. Qureshi, N. Shahzad, and G.M. Umrani; for library searches, P. Tabusa; for graphics, H. Takeuchi; and for constructive comments on an earlier draft, J.E.M. Arnold. The author alone, however, is responsible for the analysis presented here.
2. Pakistani foresters now study the vegetation in such places as the closest remaining approximation of the country's natural vegetation (Chaghtai *et al.* 1978, 1983, 1984).
3. The same argument is made concerning the deserts of contemporary western India (Gupta 1968, pp. 92–3).
4. The same definition of 'forest cover' was used in a recent study that concluded that Sri Lanka was becoming deforested—a conclusion that was reversed when the tree cover on private gardens and plantations was included in the definition (Leach 1987, pp. 44–45).
5. Fruit trees were excluded from the analysis in Table 4.1. The per centage of farmers that have planted trees would be much higher if fruit trees, the cultivation of which has a much longer history in Pakistan, were included in the analysis (cf. Cernea 1989, pp. 4–5). In other parts of the country, farmer involvement in the cultivation of forest trees reportedly is higher. Rauf (1981, p. 70) writes that 84 per cent of villagers surveyed in the Hazara district of northern Pakistan reported having planted trees.
6. The mean date (for the commencement of tree planting) of 25 years ago can be compared with the commencement of the system of agroforestry observed by

Michie (1986, p. 240) in Rajasthan, namely 30 years (thus, according to Michie's data, 1982 minus 1952 = 30 years).

7. *Junglot ke drukht*, which is a cumbersome term in Urdu, appears to be a recent development. The original opposition probably was *drukht* 'tree' versus *phal* 'fruit (tree)', with the term 'tree' thus inherently meaning '(non-fruit) tree' and, by the same virtue, the term 'fruit (tree)' meaning 'non-tree'. This also may explain the preference among Pakistani foresters (much to the consternation of the western foresters on the Forestry, Planning and Development Project) for the term 'fruity trees' as opposed to 'fruit trees': while the latter term suggests that fruit trees are merely one type of tree, the former instead implies that they partake of the qualities of something inherently non-tree-like.

8. Compare this with Shiva's (1989, p. 85) comment on the land-use category of 'wasteland' (which may have been synonymous with *jungli*):

   'Wastelands' as a land use category is, like much else, a part of our colonial heritage, loaded with the biases of colonial rule, where meaning was defined by the interest of the rulers. The colonial concept of wastelands was not an assessment of the biological productivity of land, but of its revenue generating capacity: 'wasteland' was land that did not pay any revenue because it was uncultivated.

9. See Chapter 3 for a description and analysis of such a practice, cropping rainfed pearl millet (*Pennisetum americanum*) under *khejri* (*Prosopis cineraria*), in the Rajasthan desert that borders Pakistan.

10. In the study districts, an average of 19 per cent of farm area is classified as unarable 'wasteland' (GOP 1980, Table 3.1).

11. In the study districts, an average of 53 per cent of farms are partially or completely irrigated (GOP 1980, Table 4.1).

12. In the study districts, the average household farm is fragmented into an average of 4.6 pieces (GOP 1980, Table 2.1).

13. Pakistan has the third-highest total of both large livestock (cattle, equines, buffaloes, and camels) and small livestock (goats and sheep) of the 37 nations of Asia and the Mid-East (World Resources Institute (WRI) 1990, p. 283 table 18.3).

14. The difference in emphasis is an important one. Van Ginneken (1991, p. 33) neatly sums it up as follows: 'Trying to find suitable trees to plant on the land you like [as in the FP&D Project design], is an entirely different thing, than to plant trees you like on land you're trying to find [as in the actual implementation of the FP&D Project].'

15. Thorn fences typically incorporate a wide variety of trees, bushes, and grasses, some wild, some managed, and some planted, and produce fuel, fencing, fodder and timber.

16. This critique applies largely to the traditional belief system of the Forest Service, a belief system that was targeted for change by the Forestry Planning and Development Project. The project has effected a number of major changes in forester attitude, aided by the appointment of an unusually dedicated officer to head Pakistan's Forest Service during the project's start-up phase.

17. Environmental relations that involved anything other than forest were allocated not to the Forest Department but to the completely distinct Ministry of Urban Affairs and the Environment.

18. Cernea (1985, p. 271) reports that at the time of his study in Kashmir, one out of every five households was involved in litigation with the Forest Service.

References

Agarwal, B. (1986). *Cold hearths and barren slopes: the woodfuel crisis in the third world*. Riverdale Co. for Institute of Economic Growth, Riverdale (MD).

Bartlett, H.H. (1956). Fire, primitive agriculture, and grazing in the tropics. In *Man's role in changing the face of the earth* (ed. W.L. Thomas), pp. 692–720. University of Chicago Press.

Bernier, F. (1891). *Travels in the Mogul empire: AD 1656–1668* (trans. A. Constable). S. Chand, Delhi.

Bokhari, A.S. (1989). Pakistan: a review and analysis of forest policy and legislation. World Bank Consultant's Report. World Bank, Islamabad.

Buhler, G. (trans.) (1964). *The laws of Manu*. Motilal Banarsidass, New Delhi. (Originally published 1886, Oxford University Press.)

Campbell, T. (1992). Socio-economic aspects of household fuel use in Pakistan. In *The sociology of natural resources in Pakistan and adjoining countries: case studies in applied social science* (ed. M. Dove and C. Carpenter), pp. 304–29. Vanguard Press for the Mashal Foundation, Lahore.

Castro, P. (1991). Indigenous Kikuyu agroforestry: a case study of Kirinyaga, Kenya. *Human Ecology*, **19**, 1–18.

Cernea, M.M. (1985). Alternative units of social organization sustaining afforestation strategies. In *Putting people first: sociological variables in rural development* (ed. M.M. Cernea), pp. 267–93. Oxford University Press for the World Bank, New York.

Cernea, M.M. (1989). *User groups as producers in participatory afforestation strategies*, World Bank Discussion Paper 70. World Bank, Washington, DC.

Chaghtai, S.M., Shah, H. and Akhtar, M.A. (1978). Phytosociological study of the graveyards of Peshawar District, NWFP, Pakistan. *Pakistan Journal of Botany*, **10**, 17–30.

Chaghtai, S.M., Rana, N.A., and Khattak, H.R. (1983). Phytosociology of the Muslim graveyards of Kohat Division, NWFP, Pakistan. *Pakistan Journal of Botany*, **15**, 99–108.

Chaghtai, S.M., Sadiq, A. and Shah, S.Z. (1984). Vegetation around the shrine of Ghalib Gul Baba in Khwarra-Nilab valley, NWFP, Pakistan. *Pakistan Journal of Forestry*, **34**, 145–50.

Chambers, R. and Leach, M. (1989). Trees as savings and security for the rural poor. *World Development*, **17**, 329–42.

Champion, H.G, Seth, S.K. and Khattak, G.M. (1965). *Forest types of Pakistan*. Pakistan Forest Institute, Peshawar.

Dani, A.A. (1988). Peripheral societies in a nation-state: a comparative analysis of mediating structures in development processes. Unpublished PhD thesis. University of Pennsylvania.

Dove, M.R. (1992a). Foresters' beliefs about farmers: an agenda for social science research in social forestry. *Agroforestry Systems*, **17**, 13–41.

Dove, M.R. (1992b). The dialectical history of 'jungle' in Pakistan. *Journal of Anthropological Research*, **48**, 231–53.

Dove, M.R. (1993). The coevolution of population and environment: the ecology and ideology of feedback relations in Pakistan. *Population and Environment*, **15**, 89–111.

Dove, M.R. (in press). The human ecological background of farm forestry development in Pakistan. In *Tree growing by the farmers of South Asia* (ed. V. Ballabh and N.C. Saxena). New Delhi, Sage Publications.

Ginneken, P. van. (1991). *Not seeing the people for the trees? A Review of practices and policies of tree planting.* NOVIB Report, South–North Project on Sustainable Development in Asia.

Gold, A.G. and Gujar, B.R. (1989). Of gods, trees and boundaries: divine conservation in Rajasthan. *Asian Folklore Studies*, **48**, 211–29.

GOP (1962). *The West Pakistan forest manual*, Vol. 1. GS&P Dept., Peshawar.

GOP (1980). *Pakistan census of agriculture 1980*, Vol. 2 (Parts 1, 2 and 4). Agricultural Census Organization, Statistics Division, Government of Pakistan, Lahore.

Gupta, R.K. (1968). Anthropogenic influences on the vegetation of Western Rajasthan. *Vegetatio*, **16**, 79–94.

Kangle, R.P. (1988). *The Kautilya Arthasastra*, 2nd ed. Vol. 3. Motilal Banarsidass, Delhi. (Orignally published 1969, Bombay University)

Leach, G. (1987). *Household energy in South Asia.* Elsevier Applied Science, London.

Michie, B.H. (1986). Indigenous technology and farming systems research: agroforestry in the Indian desert. In *Social sciences and farming systems research* (ed. J.R. Jones and B.J. Wallace), pp. 221–44. Westview Press, Boulder.

Mohammad, N. (1989). *Rangeland management in Pakistan.* International Center for Integrated Mountain Development (ICIMOD), Kathmandu.

Moreland, W.H. (1988). *The agrarian system of Moslem India: a historical essay with appendices.* Kanti Publications, Delhi.

Rao, S. (1957). History of our knowledge of the Indian fauna through the ages. *Journal of the Bombay Natural History Society*, **54**, 251–80.

Rauf, M.A. (1981). *Forestry development in Pakistan: a study of human perspectives.* Pakistan Forest Institute, Peshawar.

Repetto, R. (1988). *The forest for the trees? Government policies and the misuse of forest resources.* World Resources Institute, Washington, DC.

Sheikh, M.I. (1986). *A case study on hurries—Acacia nilotica block plantations for wood production in Pakistan.* Pakistan Forest Institute, Peshawar.

Shiva, V. (1989). *Staying alive: women, ecology and development.* Zed Books, London.

Supple, K.R., Razzaq, A., Saeed, I. and Sheikh, A.D. (1985). *Barani farming systems of the Punjab: constraints and opportunities for increasing productivity.* National Agricultural Research Centre, Islamabad.

Tucker, R.P. (1984). The historical context of social forestry in the Kumaon Himalaya. *Journal of Developing Areas*, **18**, 341–56.

Whyte, R.O. (1968). *Grasslands of the monsoon.* Faber & Faber, London.

World Resources Institute (1990). *World Resources 1990–1991.* Oxford University Press, New York.

Zimmermann, F. (1987). *The jungle and the aroma of meats: an ecological theme in Hindu medicine.* University of California Press, Berkeley.

5 Patterns of tree growing by farmers in eastern Africa

Katherine Warner

5.1 Introduction

With among the highest population growth rates in the world, the countries of eastern Africa have experienced rapid increases in the pressures on their land-based resources. The areas used for crop agriculture or livestock management have expanded greatly, the intensity of use of land and other resources has steadily increased in most land-use systems (LUSs), and areas of forest and woodland have been heavily reduced or changed.

However, the extent and nature of these changes have varied across the region, which encompasses marked differences in altitude, climate, soil and resource endowment. These are reflected in major differences in land use and population density. Differences in patterns of resource use also stem from variations in development of infrastructure, and in access to markets, inputs and services. Additional differences reflect the varying needs, customs and livelihood patterns of the peoples living within the region.

The present chapter draws on a recent study that examined how the patterns of tree management by farmers varied across an area of eight countries within eastern Africa (Table 5.1), and that explored how these variations related to selected physical, demographic, socio-economic and institutional characteristics of the underlying situation.[1] The aim of this cross-country assessment is to see if it provides a more complete understanding of the role of on-farm trees and tree products in different livelihood systems within the region, and of how this role evolves as farm households respond to the pressures of change on their system.

The chapter is organized around several sets of factors that are likely to affect tree management. The next section examines the effect of agroecological characteristics on land use, and on the location of trees within different patterns of land use. The following section explores the impact on the potential for tree management of the different agricultural and resource use practices that have evolved in different LUSs. The fourth section deals with tenure, and with how rights to use of land and trees influence farmer decisions about the latter. The fifth section explores the economic and social values that farm households place on tree products and services, and the choices between tree growing and alternative uses of the household's resources. The final section

Table 5.1 Countries and land area by geographical zone

Country	Population (millions)[a]	Population density (number) per km^2)	Per cent of country accounted for by land area
Bimodal Plateau Zone			
Kenya	23.5	20	15
Uganda	17.0	70	40
Rwanda	7.0	250	62
Burundi	5.0	180	85
Tanzania	22.0	30	<5
Unimodal Plateau Zone			
Malawi	8.0	65	43
Tanzania	22.0	30	41
Zambia	7.5	10	86
Zimbabwe	9.0	23	36

Notes: [a] 1990 estimates. Sources: WRI/UNEP/UNDP (1990), Hoekstra (1988, p. 2), MATF and ICRAF (1986), NTF/K and ICRAF (1988), ZATF and ICRAF (1989), Campbell (1987).

considers the effect of government policies and intervention programmes, in a region where there has been considerable intervention designed to influence local tree growing by farmers.

5.2 Tree management and agroecological characteristics

The eight contiguous eastern African countries that are examined in this chapter fall into two major geographical zones that run across the region: the Bimodal Highlands and the Unimodal Plateau (see Fig. 5.1 and Table 5.1). These zones are defined primarily on the basis of altitude, rainfall and crop growing period. They contain the high potential areas within each country, and are the areas of highest population concentration. The Bimodal Highlands zone covers Kenya, Uganda, Burundi, Rwanda, and the northern highlands of Tanzania, while the Unimodal Plateau zone covers much of the rest of Tanzania, and Malawi, Zambia and Zimbabwe. For the purposes of the study, each of the two geographical zones was divided into country specific agroecological zones, and then subdivided into land use systems.[2]

Patterns of tree growing by geographical zone

A number of features influencing land and resource use are observable across both the Bimodal Highlands and Unimodal Plateau zones. Most systems provide security of land tenure, although these may not include rights of exclusion. Throughout eastern Africa, rural households are experiencing labour shortages and a growing dependency on off-farm income, as more men seek work outside the homestead. In response to the seasonal labour shortages that result, and to increased demands for cash income, farm households are changing the 'traditional' crop repertoire. Increasing reliance is being placed on New World crops: maize, cassava, potatoes, sweet potatoes, and beans. Beer brewing, which involves heavy fuel demands, is everywhere an important non-crop activity.

Fig. 5.1 Land-use systems in the study region.

Despite the similarities, differences between the zones are readily discernible (Table 5.2). LUSs in the Bimodal Highlands zone reflect the inherent potential of higher rainfall: exhibiting the greater population density possible because families can subsist on smaller farms, involvement in valuable export crops such as coffee and tea, and (with the exception of dairy farming) less reliance on livestock. In the drier Unimodal Plateau areas, depending on more extensive agricultural systems, there is a greater dependence on livestock, and consequently, a greater emphasis placed on access to grazing land.

Farmers plant trees throughout the region, though to varying degrees and in different patterns within their farm landscapes. Trees can be established and maintained more easily in the Bimodal Highlands zone because of the two rainy seasons and a higher total rainfall. In the drier areas of the Unimodal Plateau zone, the timing of seedling planting is crucial, and farmers have to expend more time and energy in establishing and caring for seedlings. Seedling survival rates are lower in this zone, as are yields.

In addition, trees can be planted more widely in Bimodal Highlands zone areas. Because of the relatively low competition from cattle grazing, trees can be grown in crop fields as well as on boundaries and homesteads. In Unimodal Plateau areas where livestock graze freely, tree planting tends to be restricted to the protected area of the homestead.

Farmers everywhere give priority to planting fruit trees in the homestead area.[3] Even in the Chitemene LUS, where wild fruits are available and forest cover at some sites is over 70 per cent, farmers still plant exotic fruit trees in the homestead. Especially prevalent species are mango in the Unimodal Plateau zone, and avocado in the higher altitudes of the Bimodal Highlands zone. There is not much information about where farmers obtain their exotic fruit seedlings. There is generally little government support yet for planting these

Table 5.2 Zonal patterns of location and uses of trees

	Bimodal Highland Zone	Unimodal Plateau Zone
Location of trees		
Retained:	homesteads	fields and homesteads
Planted:	boundaries, fields, homesteads	homesteads
Uses of trees		
Retained:	fuel	indigenous fruit
Planted:	exotic fruits, indigenous and exotic trees (especially *Markhamia* spp., *Ficus* spp., *Euphorbia* spp., *Cassia* spp., and *Grevillea* spp.)	exotic fruits
Special concerns	competition between crops and trees	competition between livestock and trees

Table 5.3 Selected land-use systems

Bimodal Highlands	Unimodal Plateau
High Plateau	Sukuma Agropastoral
Kagera Piedmont	Chitemene
Central Congo–Nile Divide	Barotse Agropastoral
North Highlands	Maize Livestock
Lakeshore	Region III Maize Livestock
Kigezi Montane	Cereal Livestock
Coffee-based Highland	Maize Pulse
Mixed Farming	Lilongwe Plains

favoured trees, either from forestry or agricultural extension services, since the former concentrate on 'forestry' species, while the latter focus on export crops. Planting stock of exotic fruit tree species appears to come, for the most part, from farmers' own fields, from neighbours, or from non-governmental organizations (NGOs).

Farmers also favour multi-purpose trees for planting. In Unimodal Plateau areas, where livestock constitute an important component of LUSs, trees are needed which can be used for fodder as well as for poles and timber. In Bimodal Highlands areas of high population density, trees are needed for poles and construction materials, as well as for fuel and soil improvement. Trees that can supply only one of these needs, such as fuelwood, are very rarely chosen for planting.

In addition, farmers in both zones retain selected indigenous trees. Indigenous fruit trees are of importance in the Unimodal Plateau zone, being retained both in the homestead and in fields. This may be because there are many species of wild fruit to be found in miombo woodland, which is common to much of the zone. Even so, farmers tend to retain, rather than actively plant, indigenous fruit species.

It is unclear to what extent indigenous trees are planted. In some LUSs, such as the Bimodal Mixed Farming system, there is a prevalent feeling that indigenous trees just appear, or are 'God's work', so that there is no need to plant them (Wachira 1987, p. 16). Where this is the case, it could result in gradual replacement of indigenous trees by exotics.

Patterns of tree growing by LUS

Since they are delineated by distinctive environment, technology and user characteristics, there is great variation among the LUSs in each agroecological zone, as is shown in the information given in Tables 5.4–5.7, for each of the 16 selected LUSs listed in Table 5.3.[4] Patterns of tree growing by LUS in turn reveal a wide diversity of species planted and retained, and a rich variety of uses for trees in the region. Given the differences in the tree growing practices in the LUSs, as well as the variation in the data itself, uniformly comparable data were not available across the systems, but the following patterns can be discerned from the information available on the 16 LUSs listed.[5]

LUSs in the Bimodal Highlands zone   In the High Plateau LUS, trees are planted in Rwanda 'on farm boundaries, around homesteads and in the fields. Woodlots are planted on hillsides and some planting takes place on agricultural land lying fallow' (Gibson and Mueller 1987). Fruit trees are almost universal, especially avocado, but other trees are planted as well, with *Eucalyptus* spp., *Euphorbia tirucalli*, *Ficus* spp., *Markhamia platycalyx*, and *Grevillea robusta* equally common. In Burundi the same species of trees are found, but are dispersed in the fields as well as planted on the boundaries (Lamfalussy and Delince 1987).

A very large number of indigenous species are retained in the fields and near the home in some sites. The primary use of the species most commonly retained is fuelwood. This contrasts with the LUSs in the Unimodal Plateau zone, where the trees most frequently left in the fields are for fruit, rather than fuelwood.

Sector nurseries provide seedlings for the majority of exotic fruits and *Eucalyptus* spp., *Cupressus lusitanica*, and *Grevillea robusta*, but farmers also collect seedlings from natural regeneration in the fields. *Eucalyptus* woodlots are planted by farmers who have to cut on-farm trees to obtain fuelwood for the household.

This pattern of trees in the High Plateau is similar to that found in the Kagera Piedmont LUS of Rwanda, the North Highlands LUS of Tanzania and the Lakeshore LUS of Uganda. This may reflect the cropping system of intensive banana gardens that these LUSs share. In the High Plateau and Kagera Piedmont systems coffee is monocropped, a mandatory regulation of the governments, but in the Tanzania North Highlands and Ugandan Lakeshore systems coffee is integrated into the banana gardens. The composition of tree species may vary; for example, more fruit trees appear in the Lakeshore system than in the High Plateau, but the general pattern of where trees are planted and the density of trees in the farming landscape appears to be similar in each of the systems (Fernandes *et al*. 1984, 1985; Oduol and Aluma 1990).

This pattern is also found to a large extent in the Central Congo-Nile Divide LUS in Burundi, but with a marked difference in species. The trees which so often appear in eastern and southern Africa, mango, *Markhamia* spp., and *Euphorbia tirucalli*, are absent in the Divide. The Divide's extremely poor soils, its high elevation and low temperatures, etc., have resulted in a different choice of trees. Here, *Eucalyptus* spp. are the most popular trees because of their high tolerance of poor soil conditions. Fruit trees are less prevalent, and

Euphorbia tirucalli, almost universally used elsewhere as hedges, is absent (Depommier and Kaboneka 1988).

In the Kigezi Montane LUS trees are mostly found around homesteads, boundaries or in small household woodlots rather than integrated into the cropland. The per centage of households with trees is far smaller than in the Divide, but the location and species are similar. Two trees that are absent in the Divide LUS, *Euphorbia* spp. in hedges, and *Markhamia platycalyx* in fields and homesteads, are common in the Montane system (Uganda National Task Force and International Council for Research in Agroforestry (UNTF and ICRAF) 1988b).

This pattern of trees on boundaries is found in the Coffee-based Highland LUS in Kenya as well, but there is a much higher proportion of fruit trees in the Kenyan system, and the non-fruit species planted are similar to the species found in the High Plateau LUS (Minae 1988; National Task Force Kenya (NTF/K) and ICRAF 1988).

LUSs in the Unimodal Plateau zone   In the Sukuma and Barotse Agropastoral LUSs, planted trees, especially fruit trees, are found around the homesteads and, in the Barotse system, sometimes in the permanent fields as well (Tanzania Agroforestry Task Force (TATF) and ICRAF 1988a, 1988b; Barrow *et al.* 1988; Zambia Agroforestry Task Force (ZATF) and ICRAF 1989; see also, Barrow 1990). There are no trees reported planted outside these areas. Both of these LUSs, it should be noted, report shortages of fuelwood and rely heavily on cow dung and agricultural residues for fuel.

In the Chitemene, Maize Livestock, Region III Maize Livestock, and Cereal Livestock LUSs, exotic fruit trees are planted in the homestead and a large number of indigenous tree species are retained both in the fields and the homestead. The number of exotic fruit trees varies as does the species and number of the trees retained in the fields (Balderrama *et al.* 1988; ZATF and ICRAF 1988, 1989; Grundy *et al.* 1993).

In the Lilongwe LUS, this pattern of planted fruit trees in the homestead and retained trees in the fields has been expanded to include a larger number of planted species of non-fruit trees within the homestead: *Eucalyptus* spp., *Cassia siamea*, *Gmelina* spp., *Toona ciliata* and *Melia azedarach* (Malawi Agroforestry Task Force (MATF) and ICRAF 1986, 1988). Planting of *Eucalyptus* spp. has also been reported in the Zambian Maize Livestock system (ZATF and ICRAF 1989).

The trees retained in the fields in these LUSs appear to be primarily for food, usually fruit, although trees which produce leaves and fruits that can be used for animal fodder are also retained. As noted above, the miombo woodland, the common vegetation cover in much of this zone, is rich in indigenous wild fruits. Fruit trees are purposely spared when the fields are cleared, as the trees will bear fruit during the dry season when other foods may be scarce. In Zambia it was found that fruit was a significant part of the diet and was eaten by all family members (Kwesiga and Chisumpa 1990; concerning Zimbabwe, see also Campbell 1987). Not all indigenous fruit trees are equally valued, and highly favoured fruit trees are more likely to be retained. Over time this results in the woodlands surrounding the settlements gradually having a higher proportion of preferred fruit trees.

Unlike other agricultural LUSs in the Unimodal Plateau zone, trees are not retained in the fields in the Shire Highlands Maize Pulse system in Malaŵi, although exotic fruit trees and *Eucalyptus* spp., *Cassia* spp., and *Gmelina* spp. appear around the homestead (Chisimba 1987; MATF and ICRAF 1988). This is an interesting feature and at first glance makes it appear more similar to the agropastoral systems than to the other agricultural systems in the zone. However, in contrast to an agropastoral system, livestock plays a minor role in the Shire Highlands. The reason for the absence of trees in fields in the Maize Pulse system is a relatively high population density, with the smallest farms in the study area (on average 0.4 ha), fragmented into several fields, on which a series of crops are grown in an intensive system of relay cropping. This relay cropping system, which shares many characteristics with the land use systems in the Bimodal Highlands zone, utilizes the entire field for annuals with no space allotted for perennials such as trees.

5.3 Tree planting and agricultural and resource use practices

The patterns that emerge from the preceding section indicate that agroecological characteristics largely determine the agricultural practices that are possible in a given situation, and that these practices in turn have a major impact on the potential and need for tree planting. The present section explores some of the most important of these interrelationships further: the impact of the presence or absence of free ranging livestock on tree husbandry, competition between crops and trees, the compatibility of presence of trees with land preparation and cultivation practices, and the availability of supplies of needed forest products from existing natural tree stocks.

Trees and livestock management

In the anthropological aphorism 'cattle are the enemy of women', 'women' could well be substituted by 'trees'. Where free grazing is prevalent, farmers usually do not have the right to keep cattle out of harvested fields. This discourages the planting of trees in crop fields, and tree growing becomes restricted to the homesteads and other protected areas, such as the fenced fields that are maintained in the seasonal wetlands. The result is the pattern found across much of the Unimodal Plateau zone (Sukuma Agropastoral, Barotse Agropastoral, Zambian Maize Livestock, Zimbabwe Maize Livestock and Cereal Livestock LUSs) where trees are not planted outside protected areas.

In the Bimodal Highlands zone, on the other hand, livestock are either far from the fields in communal grazing areas (some sites in the High Plateau), or individual herds are small in number (Kagera Piedmont) and/or are stall-fed (North Highlands, Lakeshore, Kigezi Montane, and Coffee-based Highlands), and hedges are common. In this zone, trees can be planted outside the homestead without a large loss to livestock grazing. An exception to this is the Mixed Farming LUS in Kenya, an agropastoral system similar to those in the Unimodal Plateau zone in which herd size is relatively large, and livestock grazing has inhibited the planting (although not the retention) of trees outside the homestead or protected areas (Vonk 1983; Hoekstra 1984; Rocheleau and Hoek 1984; Wanjohi 1987).

Fodder from trees retained in fields or found in the forest can be seasonally important in the Unimodal Plateau zone, while fodder from planted multipurpose trees is of growing importance in the Bimodal Highlands zone for stall-fed livestock. However, farmers in the Unimodal Plateau zone will probably not plant trees in crop fields for fodder (or for other purposes) unless they have the right to exclude livestock from their fields.

Competition between crops and trees

Competition between crops and trees does occur with varying intensity in the different LUSs. The farmer weighs the potential gains from the tree with the possible losses from shade, water/nutrient competition, the presence of pests harboured in the trees, any restrictions the presence of trees imposes on cultivation, etc. If the negative consequences of having trees in the fields outweigh the positive ones, the trees are relocated out of the fields and into the boundaries and homestead area.

In the Unimodal Plateau zone, trees are retained in the field that produce fruit, or a good construction wood, or that have a beneficial impact on adjacent crops, or that have religious significance, or that are simply too large to remove. *Acacia albida*, for instance, is valued and retained throughout eastern Africa because it improves the yield of adjacent crops and provides shade. In the Cereal Livestock LUS in Zimbabwe, crops growing under some indigenous fruit trees (*Ficus capensis*, *Ficus burkii*, and *Paranari curatellifolia*) are recognized as having higher yields than crops elsewhere in the field (Balderrama *et al.* 1988). On the other hand, in the Malawian Maize Pulse LUS farm size is so small, and fields so fragmented, that trees are perceived as providing too much competition for the annual crops and are not found in the fields at all, not even on the boundaries.

The type of crop grown also influences decisions about inclusion of trees. In the LUSs in the Bimodal Highlands zone in which bananas are the dominant crop, trees are well integrated into the fields. The banana does well under shade, and when combined with coffee produces the mulch for coffee that is often lacking when bananas are not an integral part of the cropping system. Trees complement rather than compete in the banana–coffee garden, and have helped to create a sustainable LUS.

Field cultivation practices and the presence of trees

Depending on the particular LUS, fields are prepared and cultivated by hoe, ox plough, or tractor. The hoe is more commonly used in the Bimodal Highlands zone (seven out of eight of the LUSs), while the plough is frequently used in the Unimodal Plateau zone (six out of eight of the LUSs). Ploughing is considered to have a negative impact on tree growing on the grounds that trees have to be cleared from the field in order to be able to use the plough effectively. However, given the widespread retention of trees in the fields in the Unimodal Plateau zone, this appears to be an overstatement of what actually occurs. Even when fields are ploughed some trees are retained for shade, fruit, soil amelioration, etc.

Cultivation by tractors does encourage the removal of all trees in the fields, but of the LUSs in the study area only the North Highlands LUS in Tanzania uses tractors. (The fields which are ploughed by tractor are not the intensively worked banana–coffee home gardens, but the maize fields in the lowlands within the system, and, although trees are removed from the crop field, they are still retained and/or planted at the boundaries).

The method of field clearance and preparation can also have some impact on the number of trees retained or planted in the fields. In the Chitemene system in Zambia, trees are lopped and topped, the branches carried to another site and then burned. The cutting in one area and burning in another is an adaptation to the slow growing miombo, and allows the tree to resprout and the forest to regenerate quickly. Trees which are highly valued are not cut or burned. Such a system retains trees while transferring the nutrients held in the vegetation to the acidic leached soils of the fields (Chidumayo 1987, 1988a).

Availability of existing tree stocks

Another factor that directly influences farmers' decisions about tree planting is the availability of existing resources from which the farm household can obtain needed inputs. The amount of tree cover is dependent on the ecozone, population density and LUSs in a zone, with a general correlation between forest cover and population density.

In the Unimodal Plateau LUSs with low population density and large areas of forest cover (Chitemene, Zambian Maize Livestock, resettlement areas in Zimbabwe Maize Livestock), there is not much incentive to plant trees other than exotic fruits because the required building materials, poles, and fuel can be obtained from existing resources. Bimodal Highlands LUSs, by contrast, support areas of high population density where forests are degraded or nearly absent (High Plateau, North Highlands, Kigezi Montane, Coffee-based Highland LUSs). So farmers in these areas either have to plant trees to meet their needs, or purchase tree products, or shift to substitutes, such as bricks for housing. Tree planting consequently tends to increase as population density increases and access to forest and community reserves declines (see Table 5.8).

In particular, tree planting is stimulated by the decline in access to especially valued tree products for which there are no readily available substitutes. Thus fruit trees are usually the first trees to be planted. Although this may have initially been a response to the decline of indigenous fruit trees, exotic fruits such as mango, papaya, guava, and in the highlands avocado, have become popular and highly valued throughout the region (see for example, du Toit *et al*. 1984). On the other hand, trees are seldom planted to counter diminishing supplies of fuelwood, because this is a low value commodity and lower cost alternatives can usually be found. Even in treeless areas such as Barotse, biomass fuels such as dung and crop residue will be used, or fuelwood will be purchased, rather than trees planted specifically for the purpose (ZATF and ICRAF 1989; see also Schultz 1976).

Planting multi-purpose trees, which will supply fuelwood as one of their products, is more likely to occur when the responsibility for obtaining fuelwood for the household starts to be shared by men, as in the High Plateau, Kagera Piedmont, and North Highlands LUSs of the Bimodal Highlands zone. Even then, fuelwood has lower priority in making tree management decisions than higher valued outputs of trees for which it is less easy to find substitutes. The need for poles and construction materials, for example, is less easily met in other ways, and is therefore more likely to result in tree planting. Products such as poles and construction material are also likely to be favoured over fuelwood because they have a higher potential income value.

Similarly, planting of trees for fodder alone is not common, although existing trees are retained in the fields for that purpose. Multi-purpose trees which provide fruit, fuel, timber, *and* fodder are preferred by farmers. Trees which are planted as a source of mulch and green manure have been the focus of some major projects in the region: in Rwanda, the well documented Project Agro Pastoral (PAP) project in Nyabisindu included *Sesbania* spp., as a source of green manure (Sands 1987). Labour shortages, however, proved to be a limitation in the utilization of green manure in the fields. A similar labour constraint has been found to be a problem in cut and carry fodder for stall-fed livestock.

5.4 Tenure and rights

It has often been argued that farmers will not plant trees if there are uncertainties as to whether they will continue to have rights of access to their holdings. The implication is that farmers will not make long-term investments in their holdings unless there is the degree of security of tenure associated with freehold or private property (see Migot-Adholla *et al.* 1990).[6] However, this is not borne out by the situation in the region.

Customary versus freehold land tenure

In most of the region land is held under customary law, with ultimate ownership of the land vested in the State. The general pattern of customary land tenure is for a farmer to be given the land by a local authority, such as a village headman or council, in charge of land distribution. The farmer has 'rights of avail'; he has the land for cultivation, grazing, house site and use of natural resources. These rights may be inherited, with the particular kinship system determining if the land will be inherited from mother to daughters, father to sons, or parents to child. In areas of high population density inheritance may be limited to one rather than all of the offspring. In some of the LUSs land can also be acquired through the purchase or lease from another person.

Customary tenure is thus a usufruct system, for if the land is not farmed the rights to it are ended and the land becomes available for redistribution. But, the large number of trees found in areas of customary tenure does not support the argument that it inhibits tree planting. In Tanzania, Sukuma farmers did express some uneasiness about their tenure. However, this uneasiness was the result of the Tanzanian villagization programme that had moved entire communities from one area to another, rather than the customary land distribution system. The evidence assembled in this study indicates that, within the region, the great majority of farmers appear to feel secure about their landholdings (see Tables 5.4–5.7).

The customary land tenure system shows little sign of coming to an end; in fact, it may be spreading in some places. In Zimbabwe, for instance, commercial freehold areas are being transformed into areas for resettlement under customary arrangements.

In some areas, however, land is held under individual, private tenure, notably in Kenya. Once a title is obtained in that country, all customary rights are 'extinguished', though it appears that some degree of administrative control is maintained by the Country Councils (NTF/K and ICRAF 1988; Dewees 1991; see also Glazier 1985). The impact of the new individual private tenure system on tree planting appears to be positive. There is more tree planting on farms which have been adjudicated than on those which are not, but the primary motivation for planting trees may not be security of tenure. Initially, individual tenure does encourage tree planting since farmers plant trees to establish the boundaries needed for adjudication of the landholding under the new tenure system. As in other areas of the Bimodal Highlands, these boundaries then serve to establish rights of exclusion. Shepherd (1989) found adjudicated farms had far more trees planted on the boundaries than unadjudicated farms which planted trees in the fields.

It appears that farmers are also motivated to plant trees in these situations as the more intensive agriculture possible on privatized land encourages the removal of trees from fields to boundaries, and as access to off-farm resources is reduced as nearby areas are privatized. Brokensha and Riley (1987), for example, noted that among the Mbeere (in the Mixed Farming LUS) tree planting immediately increased after individual land title was received, but went on to note that with the communal bush disappearing farmers were becoming aware of the need to plant trees to replace those to which they no longer had access. If construction materials, fuel and fruit were not available off-farm, on-farm sources were being developed.

Distinctions between rights to use of land and trees

Traditionally, systems of customary tenure have recognized rights of use of the products of managed trees that are independent of the usufructuary rights to use the land (Fortmann and Riddell 1985; Fortmann 1987; Raintree 1987). However, as customary tenure landholdings become increasingly recognized as 'belonging' to individuals (Migot-Adholla *et al*. 1990), distinctions between tree tenure and land tenure will probably decline. Rights to trees are becoming increasingly intertwined with rights to the land on which the tree stands.

Currently, in the Bimodal Highlands zone tree and land tenure are becoming synonymous. The planter of the tree is recognized as having rights to the tree, but trees are usually only planted on land to which the planter has pre-existing rights or for which rights will be established once the tree is planted. By extension his rights to the tree may become attenuated if the planter no longer farms the land on which the tree stands.

In the Unimodal Plateau zone recognized rights to trees retained in the crop fields continue to be important, as do rights to trees in communal areas when they require special care, such as weeding, or if they are utilized for a specific purpose, such as hanging beehives. If trees are planted on cropland to which the community has access for grazing, the landholder usually does not have rights of exclusion in order to protect the seedlings, though this is changing in some communities.

Rights of exclusion

Throughout the region, the right of exclusion, specifically the rights of a farmer to exclude livestock from grazing on the household's fallow fields, appears to be more important for tree planting than the right to farm. If the farmer cannot protect seedlings from livestock grazing in the cropland, trees will only be planted where the farmer can enforce exclusion, e.g. the homestead area.

In LUSs where the livestock component is culturally and economically important, community leaders are reluctant to allow individuals to assume exclusionary rights. Fencing of cropland, for example, may be prohibited without the permission of the elders. This conflict over grazing areas will probably continue, especially since community leaders are often the owners of the larger herds. The difference in location of tree planting in the Unimodal Plateau and the Bimodal Highlands zones may be, at least partially, explained by the communal grazing rights and the importance of livestock found in the Unimodal Plateau zone.

Farmers in the Bimodal Highlands zone plant trees on the boundaries of fields, claiming the land and, perhaps more importantly, rights of exclusion. These rights of exclusion have been increasing as the resources in the area decline. As a response to (or co-evolving with) this increase in the rights of exclusion, community rights to grazing, fuelwood collection, fruit collection, etc., on individuals' land are declining even as communal areas of woodlands and grazing disappear.

5.5 Economic and social factors

Farmer decisions about tree management are influenced by a number of factors that affect the costs and benefits of doing so, and the returns relative to returns from alternative uses of the resources available to the farmer. The values that implicitly enter into such decisions are often conditioned by other considerations, including cultural beliefs and attitudes towards risk.

Availability of land, labour and capital to the farmer

Although the high population growth rates are putting increasing pressures on the relatively small amount of arable land, decline in farm size does not necessarily mean a decline in tree planting. Even though the Maize Pulse system of Malaŵi has the smallest farm size of the LUSs, trees are planted in the homestead to provide fruits, poles, construction, and fuel (MATF and ICRAF 1988). Elsewhere, situations are evolving where tree densities are increasing, or being maintained, on arable land capable of supporting crops that would generate more revenue, but require higher intensities of use of labour and capital.

Of the two, availability of labour appears to be the factor that has most impact on tree planting decisions. The outmigration of young males is a constraint on the intensification of the existing land use practices throughout the region.[7] Even where farm sizes are extremely small, as in the Malaŵian Maize Pulse system and over much of the Bimodal Highlands zone, households suffer shortages especially during seasonal peaks in demand for labour.

Tree growing can be one response to labour constraints where markets for trees products exist, such as in central Kenya,[8] because trees require low inputs and provide relatively high returns to labour. Currently, in the western highlands of Kenya the higher the population density the higher the proportion of trees in the landscape, with more productive exotic trees replacing indigenous species (Kerkhof 1990; see also van Gelder and Kerkhof 1984).

The costs of establishing and maintaining trees are low, and lack of capital is unlikely to prevent a farmer from planting trees, although it may hinder him or her in obtaining a preferred species or a large number of seedlings. Governments have established tree nurseries throughout the region, from which seedlings are usually available free during national tree planting days, even if the species selection and the number available may be limited. Farmers have also widely established their own nurseries, and have shown various other dynamic but unrecognized initiatives in obtaining tree seedlings.

On the other hand, capital is a limiting constraint for many agricultural activities. Credit is difficult for small farmers to obtain since it is usually linked to cash crops, production quotas, or the size of a landholding. Livestock generally constitute the main form of capital accumulation by households. This, plus the prestige of ownership, encourage a household to expend labour and land (in fodder crops and grazing) to maintain livestock. However, money from the sale of animals is usually not invested in agricultural improvements, but in food, school fees, medical expenses, etc. Remittances are the most common source of capital through much of the area. But there is competition within the household for this income between the capital needed for agricultural inputs and household expenses. In Zimbabwe (Region IV/V), for example, it was found that remittance money was spent on capital improvements (carts, inputs) for the farm (Balderrama *et al.* 1988).

Limited access to capital may therefore prevent households adopting more intensive agricultural options, and so indirectly encourage putting or keeping land under tree cover.[9] Trees can also be attractive as a source of capital, and of income (see Scherr and Alitsi 1990).

Access to markets for tree products

The primary motive for the planting of trees in the study area has been to achieve household self-sufficiency. Nevertheless, with improving rural road infrastructure and growing commoditization of tree products, market demand is becoming a subsidiary factor encouraging the growing of trees, especially for fruit and poles. Although the amount of income generated by the sale of fruit and fuelwood, etc., may not be large, it can play an important role in helping with the day to day household expenses and can lead to the planting of more trees (see Chapter 6).

This is likely to increase as access to markets improves, and farmers become more aware of income possibilities, in particular in the more densely populated areas where off-farm tree stocks have largely disappeared. It is encouraged by government policies that are beginning to focus on the small farmer as the producer of tree products for the nation, rather than large commercial plantations or national forest reserves. Increasingly, farmers are likely to select tree species keeping in mind their potential income value, rather than purely for their ability to supply immediate household needs.

In some areas, growing of trees by farmers as a cash crop is already well established. In the Bimodal Highlands of Kenya, for example, there is a long history of growing *Acacia mearnsii* (black wattle). This was originally planted for the sale of the wattle bark to the tannin industry. Currently, even with the decline in the wattle bark market, woodlots are maintained by many households both to meet household and market demand for charcoal and fuelwood, and as a way to keep land productive when there is a shortage of agricultural labour (Dewees 1990; also see Chapters 8 and 9).

Farm-level tree growing also may be stimulated by the needs for fuelwood to process an agricultural crop. Producers of tobacco are encouraged (or required) to grow trees to provide fuel for drying the tobacco. However, in practice, it is common to find farmers selling the trees grown on-farm for poles and construction material, and gathering or buying fuel off-farm for processing the tobacco.

Growing markets for wood fuels and poles, in fact, tend to stimulate increased exploitation of existing natural wood stocks, rather than the planting of more trees on farms. In the Unimodal Plateau zone, for example, production of charcoal for the urban market occurs in woodlands rather than on farms, thus contributing to the deforestation of large areas in the zone. Although there are legal restrictions on charcoal production in forest areas throughout the region, enforcement is difficult and usually ineffectual, because these supplies can be acquired at much less cost than farmer grown wood (see Chapter 7).

Trees, sustainability and risk management

In most LUSs farmer decisions about tree growing are found to be influenced by the advantages to be obtained from the presence of trees in controlling exposure to risk, and in improving the sustainability of the system. Boundary plantings, for example, help protect the crop field not only from livestock, but also from wind and soil erosion; trees planted in the crop field provide shade, retain moisture, and, depending on the species, may improve soil fertility through nitrogen fixation or leaf litter. If trees are available on-farm, the time and energy spent by family members in collecting fuelwood, fruit, etc., declines and more time is available for other activities.

In the Unimodal Plateau zone, the trees retained in the fields and growing in the nearby woodland assume a vital role during the dry season; wild fruit provides important nutrients for the household during a time of food scarcity, and tree fodder is used for maintaining livestock. The retained trees also benefit the soil, by both improving or maintaining fertility and inhibiting erosion (Young 1989). As natural tree stocks diminish, incorporating planted trees into the farming systems concentrates resources that had been spread throughout the landscape on to the farm, locating them where they can be protected and are easily accessible.

The multi-storied banana-coffee gardens found in the Bimodal Highlands zone are the most intensive of the integrated crop/tree systems; vertical layering increases overall productivity, and the diversity of the garden spreads labour inputs and product outputs throughout the year. The high productivity of these gardens, and the maintenance of soil fertility through mulching and composting that takes place within them, means that they are stable and sustainable.

However, like all other tree-based components of farming systems they are only part of the system; the farm cannot function with only a garden. Thus the Chagga of the North Highlands LUS of Tanzania, characterized by small homegardens and large families, have managed to maintain the stability of the system by acquiring access to land on the nearby plains. Currently, these communities with banana–coffee home gardens in the highlands and maize fields in the plains are the most prosperous in Tanzania.

The pressures of rapidly growing populations for the intensification or expansion of existing LUSs are likely to impose growing threats to some monocropping systems. Trees should play an increasing role in maintaining soil fertility and in reducing levels of risk for households and their crops in the region, and in diversifying the range of farm household products and income. This is likely to need, in addition to the present patterns of integration of trees, new niche and micro-level ways of incorporating trees (such as those described from western Kenya in Chapter 6).

Social and cultural attitudes

Tree management within the region is also influenced by a variety of cultural attitudes toward trees, some of which encourage tree retention or planting while others discourage active management. Religious traditions, for example, prohibit the Shona of Zimbabwe from cutting certain sacred species (primarily fruit trees) which are associated with the spirit guardian of the land (Wilson 1989). Such beliefs foster 'indigenous conservation' and reluctance to cut indigenous trees.

Other beliefs can create a reluctance to plant indigenous trees. In Zambia the planting of indigenous trees is culturally discouraged because it is perceived that 'only God can plant forest trees' (Kwesiga and Chisumpa 1990, p. 54). In Kenya it appears that they are viewed as 'something already there, to be used or managed' rather than to be planted (Feldstein *et al*. 1990). However, since this constraint is only on the planting of indigenous species, farmers are still free to plant exotic trees, and do so.

Societies in which trees are regarded as negative may take an adversarial stance towards trees. The Sukuma of Tanzania are reported to have such an attitude and stance (Wood 1966), combining a stated dislike for bush in distant fields (an attitude surviving from the anti-tsetse campaigns), with the felling of trees near the fields to stop the nesting of grain-eating birds. Since trees are also viewed as interfering with pasture, it might be competition with livestock production rather than diseases or birds that is the real basis of the antipathy, as the Sukuma do plant trees, especially exotic fruit trees, in their homesteads.

The cultural division of labour between men and women can also help account for the fact that farmers rarely choose to plant trees only for fuelwood, since it is traditionally women who assume responsibility for gathering fuelwood, while men are responsible for construction materials and certain cash needs of the household. Hence men place a higher priority on species which can provide income and building materials. That this is so is supported by the fact that in places where men become involved in fuelwood collection (for example, if the supply is distant, the fuelwood is transported by men with ox carts), they show greater interest in planting trees that can, among other things, provide fuelwood.

These decisions reflect the fact that planting trees implies ownership of a resource, i.e. the tree and its products and sometimes the land on which it is planted. Except for the matrilineal societies in Malawi and Zambia (and even in these societies men are given control of the land) women in the region are usually recognized as users, not owners, of resources. Permission to use a resource is determined by a woman's relationship to a male in the community, usually a father or husband. Therefore, it is the man who holds the right to make decisions about planting trees since this is an affirmation of his right to control resources in the community. Where a high proportion of rural households are now headed by women, because of the outmigration of many of the men, the complications of obtaining a decision can significantly delay the process of tree planting (see Chapter 6).

However, women as well as men can, and do, plant trees, especially fruit trees, in homestead areas. Since women are responsible for providing food for the household, a fruit tree planted in the homestead or home garden affirms a woman's rights and responsibilities, just as a tree planted in a crop field or boundary affirms a man's rights to control those resources.

5.6 External interventions

Government policies

Governments within the region have in recent years been active in launching initiatives to support farm and local level tree planting. To agricultural policy objectives of self-sufficiency in food, and the integration of animals and crops in mixed farming, have been added policies to increase production of fuelwood by farmers and, in the Bimodal Highlands zone, planting of trees for fodder and soil conservation.

These objectives shape current government programmes. However, the programmes are still burdened by attitudes inherited from past policies that created a sharp distinction between trees and agriculture. Colonial governments, though they sometimes encouraged farmers to plant woodlots as a source of income,[10] were generally against trees in the fields and against people in the forests. Farmers were told to clear all the trees from their fields, while land which formed part of grazing and fallow systems was sometimes expropriated to form forestry reserves. These attitudes lingered after independence. Farmers in Zambia, for example, continued to be told to clear all the trees from their fields and 'this tended to paint a picture of trees as something rather nasty to be removed in 'modern agriculture.' Only the planting of fruit trees was given some attention' (Gossage 1989, p. 1).

It should therefore not be surprising that many early interventions were not effective; there was little in the way of experience, or knowledge about the role of tree management at the farm level, on which to base them. Thus, interventions concerned with soil improvement through tree planting, reflecting national concerns about falling crop yields because of soil erosion, overgrazing, and fertility decline, were not successful at the farm level. Farmers recognize that soil fertility is important, but they have other pressing needs to which they give higher priority. In Burundi, for example, farmers were less interested in using trees for soil improvement and conservation, and for fodder production, than in what they perceived as their more critical need for building materials and fuelwood (Ntagunama 1989).

A similar lack of congruence between farmers needs and government policy has occurred over fuelwood projects. The apparent 'fuelwood crisis' of rapidly growing shortages generated tremendous concern throughout the region (Leach and Mearns 1988; Munslow *et al.* 1988; see also Chapter 7). Government programmes to create additional woodfuel supplies through tree planting, some quite massive in scope, were created and implemented, but most have not succeeded in their goals. The major problems with the programmes were difficulties in implementation, inappropriate species choice, and above all (as is discussed in Chapter 7), farmer disinterest because this did not coincide with their priorities or possibilities. Unfortunately the preoccupation with solving this perceived rural energy problem has distracted attention, and diverted resources, from pursuit of tree planting interventions that would be compatible with farmer objectives.

Support programmes and delivery systems

These problems have been accentuated by the institutional dichotomy of responsibilities for 'trees' and 'agriculture' that is still reflected in the government ministries and services concerned with farmers and trees.

Only Rwanda has created a truly integrated programme for agroforestry, i.e. trees integrated into the farming system. The more common pattern is for agroforestry programmes to be placed either in the agriculture ministry or in the forestry ministry or department.

Implementation of programmes focusing on household level tree planting has been hindered by the traditional orientation of such ministries. Forest ministries produce seedlings of forest species and usually do not have extension services and personnel skilled in providing services to farmers, while agricultural ministries have the extension services, but not the personnel skilled in non-fruit tree species. The result has been a lack of coordination between ministries, and insufficient skilled personnel at the community level.

Given difficulties of coordination between various ministries, and low levels of funding, problems have been encountered in providing large numbers of seedlings at a specific time to multiple sites. Poor choices of days to plant were sometimes made. In Zambia the initial date of national tree planting (World Forest Day) was at the end, not the beginning, of the rains, so that most of the seedlings died. Species choice was limited (sometimes only *Eucalyptus*) and did not reflect the agroecological zone differences in the country or the tree species that farmers were interested in planting. The government provided fuelwood-oriented tree species whereas, as has been noted throughout this review, farmers want fruit and multi-purpose trees.

This inability to engage the farmer has also been characteristic of community forestry projects. Communal woodlots were the focus of many of the programmes, yet household support was weak because it was unclear who would benefit from these programmes (Skutsch 1983). In Tanzania, for example, community forestry programmes had far lower survival rates of seedlings when compared with farm plantings (Mascarenhas *et al.* 1983; see also Skutsch 1983, 1985)

There have also been institutional problems in creating appropriate incentives for tree planting. Infrastructure may be poor, access to markets difficult and credit facilities non-existent. Lack of long-term support for a tree product also creates a problem of credibility for government programmes. Black wattle, for example, was successfully introduced in the past, but current market instability has created difficulties for producers.

The provision of incentives has concentrated so far on subsidizing establishment costs. But these do not seem necessary in order to motivate farmers to plant trees. On the contrary, government subsidies may actually have a long-term negative impact on seedling production. When seedlings are made available below cost by government nurseries, it depresses prices and discourages the creation of private nurseries. Governments will have to continue to shoulder the cost of seedling production and distribution if private nurseries are not encouraged.

Where government programmes have been successful has been in disseminating information about tree planting to the villages. Although the survival rate of trees planted on the annual tree planting day in the countries in the region is felt to be low, such programmes have been effective in getting schools, communities, and individuals involved in tree planting.

Non-governmental organizations

In some of the countries in the region, NGOs have also been very active in tree planting, and their impact is readily apparent. Many churches in the region, for example, have tree nurseries, conduct training, and actively encourage tree planting (Brokensha and Riley 1987). And some of the larger international NGOs working in the region are not only designing and implementing innovative agroforestry projects, but are also producing useful documentation. In Rwanda the CARE International Gituza project has produced an excellent series of papers concerning agroforestry project design and evaluation. Kenya has a large number of NGO agroforestry projects, one of which, the CARE Siaya project (whose activities are described in Chapter 7), is regarded as being one of the more successful projects in the region (see Scherr and Alitsi 1990).

One of the reasons for their success is that NGO projects are in many instances more responsive to the needs of community than government projects, and can usually be more flexible in their approach. The participatory approach of the CARE projects in eastern Africa, for example, has encouraged farmer interaction and flexibility in project objectives.

5.7 Future possibilities

Each country in the study area encompasses a diversity of agroecological zones and LUSs. However, a number of general patterns and features can be discerned from the material reviewed in the chapter. Overall, there is a trend towards increased tree planting as land use intensifies, with the patterns of tree management varying with agroecological characteristics, agricultural and livestock management practices, and the role of trees and tree products in livelihood strategies (see Table 5.8), as follows:

- Trees are planted in the homestead or home garden in all of the LUSs (lack of security of tenure does not seem to be a significant constraint to tree growing).
- Even when population density is low, and there is ready access to forest or woodland, trees are planted, although usually in small numbers, predominantly fruit trees near the homestead.
- Planting remains confined to the homestead, and other protected areas, where there are low population densities and extensive land-use practices involving livestock grazing or risk of fire in the fallow fields.
- As population densities increase, and access to resources off-farm decline,

there is an increase not only in the number of trees, but also in the number of tree species and planting locations in the farming system. But a point may be reached where farm size becomes so small that tree planting is again limited to the homestead.

- As access to off-farm resources declines, and rights of exclusion are strengthened, more boundary planting occurs, to protect areas from livestock, and to establish boundaries. More intensive cultivation may also encourage the relocation of trees from fields to boundaries.
- Farm woodlots appear when wood resources have to be totally met by on-farm production, or when there is a strong market for wood products, and/or when outmigration of males encourages a shift to land uses requiring less household labour (and not much capital).

In summary, farmers in the region are planting trees, although perhaps not the species, or in the numbers or locations that governments would prefer. Farmers know what trees they want, and have trees within their farming system for specific purposes. Tree planting is motivated by the needs of the household, and certain of these requirements, such as poles and fruit, are given a higher priority than others, such as fuelwood. Tree planting will continue to expand in the region in response to the decline of resources off-farm, the increase in the market for tree products, and the decline in available on-farm labour.

Support from government and non-government programmes continues to be needed in specific areas. Knowledge of tree planting and maintenance does not appear to be a constraint for the planting of trees with which farmers are familiar. However, farmers may need assistance in getting easier access to preferred species which may be difficult to propagate on farm, have a low survival rate, or for which seed stock is not available and in getting access to and bringing into use new species (see Dewees 1986). On the demand side, the need is for an improved infrastructure for marketing tree products, with access to better information on markets and market possibilities.

Government policies that support the integration of more trees on farms should recognize the different role of trees and tree products in different LUSs, and not attempt to impose a single strategy (in some LUSs there might not be a felt need for increasing tree planting). It is also important that support be provided over a long enough period to allow sustainable practices to develop. Trees require more of a time commitment than do annual crops. Programmes have to be willing to follow a tree from seedling to maturity.

Table 5.4 Land-use systems in the Bimodal Highlands zone, I

Characteristics of land-use systems	Land-use system				
	High Plateau[a]	Kagera Piedmont	Central Congo Nile Divide	North Highlands	Lakeshore
Country	Rwanda/Burundi	Rwanda	Burundi	Tanzania	Uganda
Rainfall (mm)	1000–1200	850–1000	1400–2000	750–2000	1000–2250
Altitude (m)	1550–1800	1400–1550	>1800	1000>2000	1000–2000
Vegetation	little forest	some forest	grasslands, 7 per cent woodlands	some forest	forests and grass savanna mosaic
Soils	shallow, acidic	generally poor, but soil on lower slopes and in valley bottoms is good	acidic, aluminium toxicity	fertile volcanic	deep, leached
Population, (by land-use system, 1990 estimates)	3 720 000	2 423 000	788 100	1 200 000[d]	3 000 000 (1988)
Population density (by land-use system, in number per km²)	300–500	150	155	up to 500	140
Settlement patterns	dispersed	dispersed	dispersed	densely scattered	densely scattered
Household size	5–6	6	6	10	5–10
Female headed household (per cent)	21[b], 7[c]	–	11	–	but some male outmigration

Table 5.4 (*cont.*)

			Land-use system		
Characteristics of land-use systems	High Plateau[a]	Kagera Piedmont	Central Congo Nile Divide	North Highlands	Lakeshore
Tenure security	secure	secure	secure	secure	secure
Farm size (ha)	077–2.00	2.30	0.81	home gardens: 0.68	1.00–1.50
Number of fields	2–11	2–3	14	2	home garden may have several plots
Types of fields:					
Homegarden	all	all	all	major field: banana–coffee	major field: banana coffee
Upland fields (dry)	all	all	all	majority use lowland for maize	no
Wetland cultivation	majority	rarely	majority	no	no
Land preparation	hoe	hoe	hoe	hoe in home garden; tractor or plough for maize	hoe
Households with hired labour (per cent)	25[b] 36[c]	–	56	–	large percentage hire migrant labour
Types of crops:					
Food	bananas, beans, sweet potatoes sorghum, maize	banana, beans, maize, sorghum, groundnut	peas/beans, maize, millet, sweet potato	banana, beans, taro, maize	banana, taro, yams, beans, manioc
Export/Cash	coffee	coffee, banana	tea	coffee, maize	coffee, fruit (minor)

Table 5.4 (*cont.*)

Characteristics of land-use systems	Land-use system				
	High Plateau[a]	Kagera Piedmont	Central Congo Nile Divide	North Highlands	Lakeshore
Fuelwood:					
Sources of	on-farm, gather	on-farm, crop residues, purchase	on-farm, purchase	1/2–1/4 from home garden, forest reserve, lowland fields, purchase	on-farm, swamp vegetation
Responsibility for collection	both husband and wife	both husband and wife	–	primarily husband	–
Households expressing shortages	28 per cent[b]	54–90 per cent[b]	–	minor	–
Other sources of income:					
Processed food/beer	yes	yes	no	yes	yes
Wood/tree products	no	no	no	minor	minor
Off-farm income	20 per cent of household income[b]	almost 30 per cent of household income	remittances	remittances	remittances, fishing
Per cent of households with:					
cattle >15	2	–	68	–	–
sheep/goats >50	6	–	43	–	–
Number of livestock owned:					
cattle	2	2	4	3	2–5
sheep/goats	6	4	4	2	3–5

Table 5.4 (cont.)

			Land-use system		
Characteristics of land-use systems	High Plateau[a]	Kagera Piedmont	Central Congo Nile Divide	North Highlands	Lakeshore
Livestock management:					
Communal grazing land	yes	yes	43 per cent use communal grazing: 32 per cent use communal and private grazing	no	yes, cattle
Private field grazing	yes	fallow	16 per cent only	no	no
Wetland grazing	yes	no	no	no	no
Stall-feed	no	40 per cent of households stall-feed cattle, 82 per cent stall-feed goats	no	all	yes, especially goats
Agricultural residue	fallow fields	yes, stall	yes, minor	yes	yes,
Tree fodder	negligible	yes, stall	negligible	yes	*Ficus* spp.
Fencing/hedges	yes	some sites	yes	yes	
Trees:					
Patterns	eucalypts in boundaries and woodlots; *Euphorbia* spp. in hedges; *Albizia* spp. and *Cassia spectabilis* dispersed in coffee fields; fruit trees in homestead with *Persea americana* also in fields	many fruit trees in homestead; woodlots of *Grevillea robusta* and eucalypts; *G. robusta, Gedrela serrulata* and *C. spectabilis* along roadsides; hedges of *Euphorbia tirucalli*	trees planted for timber and fuel; woodlots of eucalypts and *Acacia mearsii; P. americana* most prevalent fruit tree, also used for fuel	multispecies banana–coffee home gardens (39 species reported)	multispecies banana–coffee home gardens, for fruit, income and timber; eucalypt woodlots in swamps

Table 5.4 (*cont.*)

Characteristics of land-use systems	Land-use system				
	High Plateau[a]	Kagera Piedmont	Central Congo Nile Divide	North Highlands	Lakeshore
Preferred trees	Rwanda: *P. americana*, eucalypts, *Carica papaya*, *Grevlliea robusta*, guava	at two sites 86 per cent plant mainly *P. americana* and papaya as fruit trees	eucalypts, *Arundinaria alpina*, *Erythrina abyssinica, A. mearnsji, Ficus.* spp. *Cupressus lusitanica, P. americana*	some preference for fruit trees	strong preference for fruit trees for income

Notes: – no data available. Unless otherwise noted, percentages refer to per cent of households surveyed. [a] This is the predominant land-use system in the agroecological zone. The data utilized for the land-use system of the High Plateau is from several sites which are used to describe the prevaient land-use system. These are as follows: [b] describes the site of Gituza in Rwanda (Gibson and Mueller 1987), and [c] describes the site of Bugenyuzi in Burundi (Moussie and Muhitra 1989). [d] This is an estimate for all of Kilimanjaro. The population in the Chagga LUS is probably less, but specific data is not available.
Sources: Rwanda: ISAR and ICRAF (1988a, b), Hoekstra (1988). High Plateau: Sands (1987). Neumann and Pietrowiscz (1985). Gibson and Mueller (1987), Wanjama *et al.* (1988), Jones and Egli (1984). Kagera Piedmont: Balasubramanian and Egli (1986), Pinners and Balasubramanian (1990), Hummel (1985), Jones and Egli (1984). Burundi: ISABU and ICRAF (1988). Hoekstra (1988), Jones and Egli (1984). High Plateau: Moussie and Muhitira (1989), Lamfalussy and Delince (1987), ISABU and ICRAF (1988), Jones and Egli (1984). Central Congo/Nile Divide: Depommier and Kaboneka (1988), Ntagunama (1989), Jones and Egli (1984). Tanzania: ICRAF (1979), Fernandes *et al.* (1984), O'king'ati *et al.* (1984), O'king'ati and Mongi (1986). Uganda Lakeshore: Oduol and Aluma (1990), UNTF and ICRAF (1988b).

Table 5.5 Land-use systems in the Bimodal Highlands zone, II

Characteristics of land-use systems	Land-use system		
	Kigezi Montane	Coffee-based Highland	Bimodal Mixed Farming
Country	Uganda	Kenya	Kenya
Rainfall (mm)	1000–1500	1200–1500	850 (mean annual)
Altitude (m)	> 1800	1280–1820	1200–1300
Vegetation	savanna, forests, swamps	bushlands, grasslands	cleared acacia/combretum woodland
Soils	volcanic, fertile	moderate to high fertility	predominantely alfisols with some ultisols; erosion
Population (by land-use system, 1990 estimates)	450 000(1988)	3 887 200	1 356 000
Population density (by land-use system, in number per km²)	620	450–700	172
Settlement patterns	clustered	densely scattered	densely scattered
Household size	5	9	7
Female headed households (per cent)	–but male outmigration in some areas	–but 60 per cent of farmers are female	–but male outmigration in some areas
Tenure security	secure	secure	secure
Farm size (ha)	< 1	1.5	1.08 cropland 2.1 grazing land

Table 5.5 (cont.)

Characteristics of land-use systems	Land-use system		
	Kigezi Montane	Coffee-based Highland	Bimodal Mixed Farming
Number of fields	7	1 divided into 2 or 3 plots	cropland may be divided into plots
Types of fields:			
Home garden	all	all	all
Upland (dry)	all	all	all
Wetlands	62 per cent	–	–
Land preparation	hoe	hoe	plough
Households with hired labour (per cent)	–but common	80 per cent (for coffee)	–but common
Types of crops:			
Food	sorghum, potatoes, sweet potatoes, beans, peas, sorghum, potatoes, wheat, tobacco	maize, beans, potatoes, bananas, arrowroot	maize, beans, cow peas, pigeon peas, sweet potatoes, manioc
Export/cash		coffee, tea	cotton, sunflower (minor)
Fuelwood:			
Sources of	on-farm	on-farm, purchase	on-farm or gathered from forest, neighbour's plot
Responsibility for collection	–	female	female
Households expressing shortages	64 per cent, fuel or poles	yes, by majority	yes

Table 5.5 (*cont.*)

Characteristics of land-use systems	Land-use system		
	Kigezi Montane	Coffee-based Highland	Bimodal Mixed Farming
Other sources of income:			
Processed food/beer	–	–	–
Wood/tree products	11 per cent sold fuelwood	minor	minor
Off-farm income	remittances	remittances	82 per cent receive income from off-farm employment and remittances
Per cent of households with:			
cattle	31	83	'most'
sheep/goats	45	77	'most'
Number of livestock owned:			
cattle	few	2–3	7.6 (average)
sheep/goats	few	2	10
Livestock management:			
Communal grazing land	yes, all	no	no
Private field grazing	27 per cent	no	yes
Wetland grazing	yes	no	no
Stall-feed	12 per cent goats	all	few
Agricultural residue	–	yes	oxen
Tree fodder	–	minor	minor
Fencing/hedges	yes	yes	yes

Table 5.5 (*cont.*)

	Land-use system		
Characteristics of land-use systems	Kigezi Montane	Coffee-based Highland	Bimodal Mixed Farming
Trees			
Patterns	70 per cent of households planted eucalypts in woodlots and/or homestead; trees used primarily for construction, fuel, boundaries and fences	fruit trees in homestead; *Croton megalocarpus*, *Markhamia lutes*, *Ficus* spp. and *Euphorbia* spp. in boundaries; *G. robusta* in boundaries and fields	trees retained in fields; fruit trees planted in fields and homestead; *Euphorbia* spp., *Combretum* spp. and *Acacia* spp. planted on boundaries; woodlots are common
Preferred trees	eucalypts, *E. abyssinica*, *A. mearnsii*, *C. lusitanica*, Euphorbia spp.	*G. robusta*, fruit trees, *C. lusitanica*	*Acacia* spp., with fruit trees of increasing importance (especially citrus and mango)

Notes: – no data available. Unless otherwise noted, percentages refer to per cent of households surveyed.
Sources: Kigezi Montane, Uganda: UNTF and ICRAF (1988a). Coffee Based Highland, Kenya: Minae (1988), NTF/K and ICRAF (1988), Scherr and Alitsi (1990). Mixed Farming, Kenya: Rocheleau and Hoek (1984), Vonk (1983), Rocheleau (1984), Hoekstra (1984), Wanjohi (1987), Wachira (1987).

Table 5.6 Land-use systems in the Unimodal Plateau zone, I

Characteristics of land-use systems	Land-use system				
	Sukuma Agropastoral	Chitemene	Barotse Agropastoral	Maize Livestock	Region III Maize Livestock
Country	Tanzania	Zambia	Zambia	Zambia	Zimbabwe
Rainfall (mm)	600–700	1000–500	600–1000	800–1000	600–700
Altitude (m)	100–1500	1200 (average)	800–1200	800–1400	900–1200
Vegetation	almost treeless	wetter miombo	miombo, grasslands	miombo, acacia	dry miombo
Soils	varies, mostly clayish	sandy loam, leached, acidic	infertile sand, alluvial valleys	relatively fertile	shallow, moderately leached, low fertility
Population, (by land-use system, 1990 estimates)	3 427 100	400 000 (est.)	610 000 (est.)	1 500 000	2000[a]
Population density (by land-use system, in number per km^2)	35–50	2–4	4	11	35
Settlement patterns	village	dispersed	densely scattered	villages	villages
Household size	6	5–9	5–9	5–9	5
Female headed households (per cent)	some outmigration of males	30–40	large numbers of males migrate	large numbers of males migrate	outmigration of males is not allowed
Tenure security	land secure; trees belong to planter	land secure; trees belong to planter	land secure; community grazing rights	usufruct, must ask permission to fence	site of resettlement community secure

Table 5.6 (*cont.*)

Characteristics of land-use systems	Land-use system				
	Sukuma Agropastoral	Chitemene	Barotse Agropastoral	Maize Livestock	Region III Maize Livestock
Farm size (ha)	3	1	1–5	1.5–3	5, plus grazing rights
Number of fields	several fields and/or plots	several fields in crop sequence	several fields	several fields	several fields
Types of fields:					
Home garden	yes, all	yes, all	yes	yes	yes
Upland fields (dry)	yes	yes	yes	yes	yes
Wetland cultivation	minority (rice)	no	yes, most	yes	yes
Land preparation	plough	slash/burn	hoe/plough	plough	plough/hoe
Households with hired labour (per cent)	–	hire for male tasks if have funds	–	–	–
Types of crops:					
Food	maize, sorghum, millet, cassava	millet, beans, cassava, groundnuts	maize, cassava, sorghum, millet	maize, groundnuts	maize
Export/Cash	cotton, maize, rice	maize	maize	maize, cotton, sunflower, tobacco	maize
Fuelwood: Sources of	gathered, purchased, crop residues, cow dung	gathered, purchased	cow dung, crop residues, gathered, purchased	gathered, cut live trees	gathered

Table 5.6 (*cont.*)

	Land-use system				
Characteristics of land-use systems	Sukuma Agropastoral	Chitemene	Barotse Agropastoral	Maize Livestock	Region III Maize Livestock
Fuelwood (*cont.*)					
Responsibility for collection	female	male if distant, female if closer	men (28 per cent) men cut trees	men (40 per cent)	primarily female
Households expressing shortages	yes	no	yes	yes	no
Other sources or income:					
Processed food/beer	yes	yes	yes	yes	yes
Wood/tree products	no	no	no	minor	no
Off-farm income	remittance	remittance	remittance	remittance	–
Per cent of households with: cattle	70	Bemba do not keep livestock, other groups to the south keep small numbers	50	30	yes, but no data about populations
sheep/goats	–		–	> 35	
Number of livestock owned: cattle	10–50	–	10–15	4–8	–
sheep/goats	–	–	–	–	–
Livestock management:					
Communal grazing land	yes	–	yes	yes	yes
Private field grazing	fallow	–	yes, fallow	yes, fallow	fallow
Wetland grazing	fallow	–	yes, fallow	yes, fallow	none

Table 5.6 (*cont.*)

Characteristics of land-use systems	Sukuma Agropastoral	Chitemene	Barotse Agropastoral	Maize Livestock	Region III Maize Livestock
Livestock management (*cont.*)					
Stall-feed	no	–	no	no	no
Agricultural residue	fallow	–	fallow	fallow	fallow
Tree fodder	minor	–	minor	minor	–
Fencing/hedges	homestead	–	no	yes, wetlands	yes, thorny fences
Trees					
Patterns	trees planted at homestead, roadsides, small woodlots, fences; grazing inhibits tree growing in fields	*Pterocarpus angolensis* and wild fruit trees retained in fields; other indigenous trees at homestead	fruit and nut trees planted in fields and homestead	*Acacia* spp., wild fruits and *P. angolensis* retained; exotic fruits and *Eucalyptus grandis* planted in homestead and fields	trees, especially fruit trees, retained in fields; fruit and other trees planted at homestead
Preferred trees	fruit and multipurpose trees	strong preference for fruit trees, both retained and planted	fruit and multi-purpose trees	fruit trees	fruit trees

Land-use system

Notes: – no data available. Unless otherwise noted, percentages refer to per cent of households surveyed. [a]Site-specific population data of resettlement community.

Sources: Sukuma Agropastoral, Tanzania: TATF and ICRAF (1988a,b), Barrow *et al.* (1988), Mascarenhas *et al.* (1989), Kikula *et al.* (1983), Wood (1966). Chitemene, Zambia: ZATF and ICRAF (1988, 1989), Mansfield (1976a,b), Chidumayo (1987, 1988b), Holden (1988), Schultz (1976), Stolen (1983), Vedeld (1983). Barotse Agropastoral, Zambia: ZATF and ICRAF (1989), Chidumayo (1988b), Schultz (1976). Maize Livestock, Zambia: ZATF and ICRAF (1989), Schultz (1976), Kwesiga and Chisumpa (1990), Gossage (1989), Bohlin and Larsson (1983). Region III, Maize Livestock, Zimbabwe: Grundy *et al.* (1993).

Table 5.7 Land-use systems in the Unimodal Plateau zone, II

Characteristics of land-use systems	Land-use system		
	Region IV Cereal Livestock	Shire Highlands Maize Pulse System	Lilongwe
Country	Zimbabwe	Malawi	Malawi
Rainfall (mm)	200–1100 (average 559)	800–1300	800–1200
Altitude (m)	550–750	800–1300	1000–1200
Vegetation	mopane, miombo, uapaca woodlands	grasslands, isolated trees	acacia woodland
Soils	variable, shallow/heavy	moderate to high soil fertility	fairly fertile
Population (by land-use system, 1990 estimates)	143 450[a]	1 460 000	788 000
Population density (by land-use system, in number per km²)	55	> 150 up to 250	114
Settlement pattern	dispersed clusters	villages	villages
Household size	5	5	5
Female headed households (per cent)	18 50 per cent of males aged 20–29 years are absent	seasonal outmigration of males	seasonal outmigration of males
Tenure security	secure	secure if not on estate; if on estate, usufruct only	secure
Farm size (ha)	2.8	0.4	1.5

Table 5.7 (cont.)

Characteristics of land-use systems	Land-use system		
	Region IV Cereal Livestock	Shire Highlands Maize Pulse System	Lilongwe
Number of fields	several	upland field divided into 2 or 3 plots	several fields, plots
Types of fields:			
Homegarden	yes	yes	yes, usually vegetable
Upland (dry)	yes	yes	yes, all
Wetlands	26 per cent maize field 44 per cent vegetables	yes	yes, all
Land preparation	plough	hoe	hoe, but 5 per cent use plough
Households with hired labour (per cent)	—	few households can afford	hired during peak labour demand periods
Crops:			
Food	maize, millet sorghum, groundnut	maize, cassava, banana, cowpeas, pigeon peas, sorghum, rice	maize, groundnuts, beans, cassava, sweet potatoes
Export/Cash	maize, vegetables	tobacco, chickpeas	tobacco, maize, groundnuts
Fuelwood:			
Sources of	gather	crop residues, grown on farm	gathered, purchased, crop residues
Responsibility for collection	primarily female	primarily female	primarily female
Households expressing shortages	yes	yes	yes

Table 5.7 (*cont.*)

Characteristics of land-use systems	Land-use system		
	Region IV Cereal Livestock	Shire Highlands Maize Pulse System	Lilongwe
Other sources of income:			
Processed food/beer	yes	yes	yes
Wood/tree products	no	yes	yes
Off-farm income	remittances and off-farm income provide 60.5 per cent of income	remittances	seasonal remittances
Per cent of households with:			
cattle	75	4.5	16
sheep/goats	> 65	small number of households keep goats for rituals	42
Number of livestock owned:			
cattle	65 per cent of households own less than 8	–	–
sheep/goats	–	–	–
Livestock management:			
Communal grazing land	yes	yes, wetlands	yes, wetlands
Private field grazing	fallow	fallow	fallow
Wet land grazing	fallow	yes	yes
Stall-feed	no	goats	no
Agricultural residue	fallow	cattle use fallow, goats are stall-fed	yes
Tree fodder	yes, 41 per cent cut tree foliage in dry season	minor	yes, browse
Fencing/hedges	no	–	sisal

Table 5.7 (*cont.*)

Characteristics of land-use systems	Land-use system		
	Region IV Cereal Livestock	Shire Highlands Maize Pulse System	Lilongwe
Trees:			
Patterns	trees, especially fruit, retained in fields and planted at homestead: trees a notable source of fodder	trees planted at homestead: fruit, eucalypts, *Gmelina arborea* and *Cassia siamea*; mango a significant food.	trees retained in fields, especially for fodder and mulch; trees planted mainly at homestead mango also in fields mango, also eucalypts, *C. siamea*, *G. arborea*, *Toonaciliata* and *Melia azedarach*
Preferred trees	fruit trees and multipurpose trees for fodder	fruit trees, and *Cassia*, *Gmelina* and *Eucalyptus* spp.	

Notes: – no data available. Unless otherwise noted, percentages refer to per cent of households surveyed. [a] Population data are based on one site. The total population of the LUS is considerably higher.

Sources: Region IV, Cereal Livestock, Zimbabwe: Balderrama *et al.* (1988), Wilson (1989). Shire Highlands, Maize Pulse system, Malawi: MATF and ICRAF (1986), FAO (1985), ESU (1981), McCall and Skutsch (1987), Chisimba (1987), Chandele *et al.* (1987), Manda *et al.* (1985). Lilongwe, Malawi: MATF and ICRAF (1988), MATF and ICRAF (1986).

Table 5.8 Trends associated with a decline in access to off-farm tree resources

Characteristics of tree and land-use management	Changes which can be anticipated with declining access to off-farm tree resources
Land-use system	
Tree tenure	Bimodal: tree and land tenure become increasingly intertwined Unimodal: continued rights to trees retained in fields
Right of exclusion	Bimodal: control of access to crop fields increases Unimodal: control of livestock access to fallow fields still maintained
Forest/woodland system	Decrease in gathered forest and wood products
Farm size	Decreases in size, increases in fragmentation
Field management	Fallow decreases. Final stage is relay cropping
Livestock management	Bimodal: livestock increasingly stall-fed or eliminated Unimodal: increasing reliance on fallowed fields for grazing and on trees used for fodder
Trees	
Location in landscape	Concentration of tree resources shifts from off-farm to on-farm
Retained trees	Gradual supplementation and/or replacement of retained indigenous trees with planted exotic species
Planted trees	Initially homestead, then expansion of niches to fields and field boundaries[a]
Planted species	Initially fruit trees then expansion to include poles and timber species
Number of trees planted	Generally, increases in response to decline of access to off-farm tree resources

Table 5.8 (*cont.*)

Characteristics of tree and land-use management	Changes which can be anticipated with declining access to off-farm tree resources
Tree products	
Fuelwood	Needs increasingly met by on-farm substitutes (such as crop residues) and by more intensive management of on-farm tree resources
Timber/poles/construction	Needs increasingly met by planting trees
Fodder	Bimodal: needs increasingly met by cut and carry systems, but labour availability is the limiting factor in livestock management
	Unimodal: reliance on grazing areas and fallowed fields; livestock is phased out in some land-use systems

Notes: [a] However, in the Unimodal Maize Pulse LUS, the small farm size combined with relay planting has resulted in trees being primarily planted in the homestead.

Notes

1. The information in this chapter is based on a joint Oxford Forestry Institute (OFI) and ICRAF study (Warner 1993). This was based on information obtained from the literature, from unpublished records, and from discussion with researchers working in this and related areas in each of the eight countries. The research was funded by the Rockefeller Foundation. The author wishes to express her appreciation for the help of OFI and ICRAF staff, and members of the country teams within the Agroforestry Research Network for Africa (AFRENA) network and others working in the eight countries, who provided unpublished documents and expertise, helped organize visits to the countries, and commented on the country drafts. The principal collaborators in the parent institutions were Sara Scherr at ICRAF and J.E.M. Arnold at OFI. Dawit Seyoum, an OFI Research Fellow funded by the Rockefeller Foundation, aided in compiling data from the ICRAF library and in collecting information in some of the countries.
2. The zonal delineation was based on that developed by ICRAF, with its national collaborators, in the course of the AFRENA country studies to define agroforestry potentials and research needs. The **Bimodal Highland** zone is defined as having:
   - altitude range of 1000–2500 m above sea level;
   - rainfall of at least 1000 mm per year, in two rainy seasons;
   - crop growing period of 90–270 days;
     in some areas of the zone (e.g. Burundi and Rwanda) very little natural vegetation remains; where natural vegetation does occur it is usually forest, although the particular species composition varies.
   The **Unimodal Plateau** zone is defined as follows:
   - altitude range of 600–1600 m;
   - rainfall is in a unimodal pattern 600–1500 mm;
     crop growing period of 90–270 days; the 180 growing day isoline was used to further delineate the wetter and drought prone areas;
   - vegetation consists mainly of miombo (*Brachystegia*, *Julbernardia*, and *Isoberlinia* are the major species) or Acacia woodlands (*Acacia*, *Piliostigma*, and *Combretum*).
   The defining characteristics of the selected LUSs are shown in Tables 5.4 and 5.5.
3. The only exception to this preference for fruit trees was found in Burundi, in the Central Congo–Nile Divide LUS, where farmers want more trees for construction, fuel and timber.
4. The coverage of the study was shaped by data limitations and the large number of zones and LUSs. Decisions as to which agroecological zones and LUSs to include in the analysis were based on the importance of the LUS to the country (focus area for projects, economic importance, size of population, etc.) and the quality of the data. The LUSs represented in Tables 5.4–5.7 are the major systems for which data were available on the farming system, household characteristics, environment and the local economy. With the exception of Rwanda and Burundi, each country is represented in the tables by at least two LUSs. The 16 systems documented contain more than 25 per cent of the total population in the eight-country region (Warner 1993).
5. The detailed data for the individual LUSs are found in Warner (1993, table 2.6–2.24).
6. It has also been argued that lack of individual or freehold tenure will limit access to credit, and therefore discourage tree planting as well as other investments. This

has not proven to be a very important factor since farmers are extremely reluctant to use land as collateral for loans. Also, there is little credit available to the small-holder in the region, and what credit does exist is usually tied to cash crop inputs. To obtain credit for cash crops the farmer does not have to use his land as collateral, although he usually has to be a large enough producer to qualify for a loan.

7. In both the Unimodal Plateau and Bimodal Highland zones the shortage of labour also reduces the likelihood that agricultural residues and dung will be worked into the soil, which in some circumstances can increase the fuel supplies that are available to the household.

8. See Chapter 9 for a detailed discussion of land and labour allocation and tree growing, illustrated from the situation in an area of central Kenya.

9. However, trees rank low in the allocation of limited amounts of capital available for agricultural inputs. Minae (1988) has reported that, among the crops grown in the Bimodal Highlands of Kenya, trees are given the lowest priority in terms of resource allocation and management inputs.

10. For example, the black wattle woodlots in central Kenya that are discussed in Chapters 8 and 9, and *Eucalyptus* woodlots in Rwanda.

References

Balasubramanian, V. and Egli, A. (1986). The role of agroforestry in the farming systems in Rwanda with special reference to the Bugesera–Gisaka–Migongo (BGM) region. *Agroforestry Systems*, **4**, 271–89.

Balderrama, S., Fenta, T., Hussein, A., Jackson, C., Midre, M., Scott, C., Vasquez, C. and de Vos, J. (1988). *Farming systems dynamics and risk in a low potential area: Chivi South, Masvingo Province, Zimbabwe*, ICRA Bulletin 27. International Course for Development Oriented Research in Agriculture, Wageningen.

Barrow, E. (1990). *Evaluating the effectiveness of participatory agroforestry extension programmes in a pastoral system, based on existing traditional values: a case study of the Turkana in Kenya*. Paper presented at a workshop on participatory methods for on-farm agroforestry research, held at the International Council for Research in Agroforestry, Nairobi, Kenya, 19–23 February 1990.

Barrow, E., Kabelele, M., Kikula, I. and Brandstrom, P. (1988). Soil conservation and afforestation in Shinyanga region: potentials and constraints. Unpublished report to NORAD. Nairobi.

Bohlin, F. and Larsson, F. (1983). *Village tree plantation study carried out for IRDP, Eastern Province, Zambia*, International Rural Development Centre Working Paper, Swedish University of Agricultural Sciences, Lund.

Brokensha, D. and Riley, B.W. (1987). Privatization of land and tree planting in Mbeere, Kenya. In *Land, trees and tenure*, Proceedings of an International Workshop on tenure issues in agroforestry held in Nairobi, Kenya, 27–31 May 1985, (ed. J.B. Raintree), pp. 187–92. ICRAF and the Land Tenure Center, Nairobi and Madison.

Campbell, B.M. (1987). The use of wild fruits in Zimbabwe. *Economic Botany*, **41**, 375–85.

Chandele *et al.* (1987). *A study of the smallholder farming system in West Mulange, Malawi*, ICRA Bulletin 24. International Course for Development Oriented Research in Agriculture, Wageningen.

Chidumayo, E.N. (1987). A shifting cultivation land use system under population pressure in Zambia. *Agroforestry Systems*, **5**, 15–25.

Chidumayo, E.N. (1988a). Integration and role of planted trees in a bush-fallow cultivation system in central Zambia. *Agroforestry Systems*, **7**, 63–76.

Chidumayo, E.N. (1988b). Village tree planting in Zambia: problems and prospects. *Desertification Control Bulletin*, **17**, 22–6.

Chisimba, L. (1987). Rural energy use and afforestation survey in Malaŵi. Unpublished paper. Department of Social Statistics, University of Southampton.

Depommier, D. and Kaboneka, S. (1988). *Analyse de l'enquête agroforestiere realisée dans la commune de Gisozi Burundi. Hautes terres d'Afrique de l'est à regime pluviométrique bimodal*. ICRAF, Nairobi.

Dewees, P.A. (1986). Economic issues and farm forestry. Unpublished paper prepared for the World Bank Kenya Forestry Sector Review. Nairobi.

Dewees, P.A. (1990). *Why rural people take care of trees*. Oxford Forestry Institute, Oxford.

Dewees, P.A. (1991). *Woodlots, labour use and farming systems in Kenya*. Paper presented at the Institute of Rural Management workshop on socio-economic aspects of tree growing by farmers in South Asia at Anand, India, March, 1991.

ESU (Energy Studies Unit) (1981). *Malaŵi Smallholder Tree-Planting Survey*. Ministry of Agriculture, Lilongwe, Malaŵi.

FAO (Food and Agriculture Organization) (1985). *Understanding tree use in farming systems*. Paper presented at the Workshop on planning fuelwood projects with participation of rural people, held in Lilongwe, Malawi, 12–30 November 1984.

Feldstein, H.S., Rocheleau, D. and Buck, L.E. (1990). Agroforestry extension and research: a case study from Siaya District. In *Working together: gender analysis in agriculture*, Vol. 1, (ed. H.S. Feldstein and S.V. Poats), pp. 167–208. Kumarian Press, West Hartford.

Fernandes, E.C.M., O'kting'ati, A. and Maghembe, J. (1984). The Chagga home gardens: a multistoried agro-forestry cropping system. *Agroforestry Systems*, **2**, 73–86.

Fernandes, E.C.M., O'kting'ati, A. and Maghembe, J. (1985). The Chagga home gardens: a multistoried agro-forestry cropping system on Mt Kilimanjaro, northern Tanzania. *Food and Nutrition Bulletin*, **7**, 29–36.

Fortmann, L. (1987). Tree tenure: an analytical framework for agroforestry projects. In *Land, trees and tenure*, Proceedings of an International Workshop on tenure issues in agroforestry held in Nairobi, Kenya, 27–31 May 1985, (ed. J.B. Raintree), pp. 17–34. ICRAF and the Land Tenure Center, Nairobi and Madison.

Fortmann, L. and Riddell, J. (1985). *Trees and tenure*. Land Tenure Center and ICRAF, Madison, Wisconsin and Nairobi.

Gelder, B. van and Kerkhof, P. (1984). *Agroforestry survey in Kakamega District*. Unpublished report to the KWDP (Kenya Woodfuel Development Project). Beijer Institute, Nairobi.

Gibson, D.C. and Mueller, E.U. (1987). *Diagnostic surveys and management information systems in agroforestry project implementation: a case study from Rwanda*. ICRAF Working Paper No. 49. ICRAF, Nairobi.

Glazier, J. (1985). *Land and the uses of tradition among the Mbeere of Kenya*. University Press of America, Lanham.

Gossage, S.J. (1989). *Agroforestry extension experiences from the national soil conservation and agroforestry extension programme in Zambia with particular reference to Eastern Province*. Paper presented at the First Zambian National Agroforestry Workshop at Lusaka, Zambia, 16–19 April 1989.

Grundy, I.M., Campbell, B.M., Balebereho, R., Cunliffe, R., Tafangenyasha, C.,

Fergusson, R. and Parry, D. (1993). Availability and use of trees in Mutanda Resettlement Area, Zimbabwe. *Forest Ecology and Management*, **56**, 243–66.

Hoekstra, D. (1984). *Agroforestry systems for the semiarid areas of Machakos District, Kenya*. ICRAF, Nairobi.

Hoekstra, D. (1988). *Summary of zonal agroforestry potentials and research across land-use systems in the highlands of eastern and central Africa*, AFRENA Paper No. 15. ICRAF, Nairobi.

Holden, S. (1988). *Farming systems and household economy in New Chambeshi, Old Chambeshi and Yunge villages near Kasama, Northern Province, Zambia: an agroforestry baseline study*. Agricultural University of Norway, Ås.

Hummel, L. (1985). *Etude sur les systèmes de production agricole et le rôle des arbres dans l'exploitation agricole: le cas de la commune de Muyaga*. Communications du Départment de Foresterie No. 4., Kigali, Rwanda.

ICRAF (1979). Tree based farming systems for some ecological regions of Tanzania. Unpublished research project document. ICRAF, Nairobi.

ISABU and ICRAF (1988). *Potentiel agroforestier des systèmes d'utilisation des sols des hautes terres d'Afrique de l'est à regime pluviométrique bimodal: Burundi*, AFRENA Paper No. 2. ICRAF, Nairobi.

ISAR and ICRAF (1988a). *Potential agroforestier dans les systèmes d'utilisation des soîs des hautes terres d'Afrique de l'est à régime pluviométrique bimodal: Rwanda*, AFRENA Paper No. 1. ICRAF, Nairobi.

ISAR and ICRAF (1988b). *Propositions de recherche agroforestiere pour le systeme de plateaux et collines au Rwanda*, AFRENA Paper No. 13. ICRAF, Nairobi.

Jones, W.I. and Egli, R. (1984). *Farming systems in Africa: the Great Lakes Highlands of Zaire, Rwanda, and Burundi*, Technical Paper No. 27. World Bank, Washington, DC.

Kerkhof, P. (1990). *Agroforestry in Africa: a survey of project experience*. Panos Institute, London.

Kikula, I., Mascarenhas, A. and Nilsson, P. (1983). *Report to support village afforestation in Tanzania*. Institute of Resource Assessment, University of Dar es Salaam, Dar es Salaam.

Kwesiga, F. and Chisumpa, S.M. (1990). Ethnobotanical survey in Eastern Province, Zambia. Draft report. ICRAF, Nairobi.

Lamfalussy, L. and Delince, J. (1987). *Enquête agro-forestiere realisée dans la province de Muyinga*. Project Reboisement Banque Mondiale-Fac., Département des Eau et Forêts, République du Burundi, Bujumbura.

Leach, G. and Mearns, R. (1988). *Beyond the woodfuel crisis: people, land and trees in Africa*. Earthscan, London.

McCall, M. and Skutsch, M. (1987). *Malaŵi woodfuel energy study: current situation and prospects*, Occasional Paper No. 5. Technology and Development Group, University of Twente, Twente.

Manda, D.R.B., Dzowela, B.H. and Johnson, W.H. (1985). *Agricultural research resource assessment in the SADCC countries*, Vol. 2 (Country report Malaŵi). DEVRES for the United States Agency for International Development.

Mansfield, J.E. (1976a). *Land resources of the Northern and Luapula Provinces, Zambia: a reconnaissance assessment*, Vol. 2 (Current land use), Land Resource Study No. 19, Lusaka.

Mansfield, J.E. (1976b). *Land resources of the Northern and Luapula provinces, Zambia: a reconnaissance assessment*, Vol. 5 (Social and economic factors), Land Resource Study No. 19, Lusaka.

Mascarenhas, A., Kikula, I., and Nilsson, P. (1983). *Support to village afforestation in Tanzania*. Institute of Resource Assessment, University of Dar es Salaam, Dar es Salaam.

MATF and ICRAF (1986). *A blueprint for agroforestry research in the unimodal upland plateau of Malaŵi*, AFRENA Paper No. 5. ICRAF, Nairobi.

MATF and ICRAF (1988). *Agroforestry research project proposal for the Lilongwe land-use system*, AFRENA Paper No. 8. ICRAF, Nairobi.

Migot-Adholla, S., Hazell, P.B., Blarel, B., and Place, F. (1990). *Land tenure reform and agricultural development in SubSaharan Africa*. Paper given at a seminar on land issues and agricultural productivity in Kenya, held at the World Bank Eastern and Southern Africa Regional Office, Nairobi, Kenya, 19 April 1990.

Minae, S. (1988). *Agroforestry research project proposal for the coffee based system in the bimodal highlands, central and eastern provinces, Kenya*, AFRENA Paper No. 16. ICRAF, Nairobi.

Moussie, M. and Muhitira, C. (1989). *An overview on the general characteristics of the farm: Bugenyuzi commune, Karuzi*. ISABU/SFSR, Burundi.

Munslow, B., Katerere,Y., Ferf, A. and O'Keefe, P. (1988). *The fuelwood trap: a study of the SADCC region*. Earthscan, London.

Neumann, I. and Pietrowicz, P. (1985). *Agroforesterie à Nyabisindu: recherches sur l-intégration des arbres et des haies dans l'agriculture*. Projet Agro-Pastoral de Nyabisindu, Etudes et Expériences, No. 9. Nyabisindu, Rwanda.

Ntagunama, F. (1989). *Projet C.V.H.A.: Cultures villageoises de haute altitude*. Paper presented at the Seminaire National l'Agroforesterie au Burundi, held in Bujumbura, 28–31 March 1989. ISABU, Burundi.

NTF/K and ICRAF (1988). *Agroforestry potentials for the land use systems in the bimodal highlands of eastern Africa, Kenya*, AFRENA Paper No. 3. ICRAF, Nairobi.

Oduol, P.A. and Aluma, J.R.W. (1990). The banana (*Musa* spp.)—Coffee robusta: traditional agroforestry system of Uganda. *Agroforestry Systems*, **11**, 213–26.

O'king'ati, A. and Mongi, H.O. (1986). Agroforestry and the small farmer: a case study of Kilema and Kirua Vunjo in Kilimanjaro. *International Tree Crops Journal*, **3**, 257–65.

O'kting'ati, A., Maghembe, J., Fernandes, E.C.M. and Weaver, G.H. (1984). Plant species in the Kilimanjaro agroforestry system. *Agroforestry Systems*, **2**, 177–86.

Pinners, E. and Balasubramanian, V. (1990). *Evolution of need-based agroforestry systems in the Bugesera and Gisaka-Migongo regions of Rwanda: Results of a farmers survey*. Paper presented at a workshop on participatory methods for on-farm agroforestry research, held at the International Council for Research in Agroforestry, Nairobi, Kenya, 19–23 February 1990.

Raintree, J.B. (ed.) (1987). *Land, trees and tenure*, Proceedings of an International Workshop on tenure Issues in agroforestry held in Nairobi, Kenya, 27–31 May 1985. ICRAF and the Land Tenure Center, Nairobi and Madison.

Rocheleau, D. (1984). *Criteria for re-appraisal and re-design: intra-household and between household aspects of FSRE in three Kenyan agroforestry projects*. Paper presented at the annual farming systems research and extension symposium, held at Kansas State University, Manhattan, Kansas, 7–10 October 1984.

Rocheleau, D. and Hoek, A. van den (1984). *The application of ecosystems and landscape analysis in agroforestry diagnosis and design: a case study from Kathama sub-location, Machakos District, Kenya*, Case Studies in Agroforestry Diagnosis and Design No. 4. ICRAF, Nairobi.

Sands, M. (1987). Integrated soil regeneration in Rwanda. In *Experiences in success*, (ed. K. Tull, M. Sands, and M. Altieri), pp. 32–9. Rodale International, Emmaus, PA.

Scherr, S. and Alitsi, E. (1990). *The development impact of the CARE-Kenya Agroforestry Extension Project*, Draft Report. ICRAF and CARE-Kenya, Nairobi.

Schultz, J. (1976). *Land use in Zambia, part I: the basically traditional land use systems and their regions*, Afrika-Studien Nr. 95. Weltforum Verlag, Munich.

Shepherd, G. (1989). Assessing farmers' tree-use and tree-planting priorities: a report to guide the ODA/Government of Kenya Embu–Meru–Isiolo Forestry Project. Overseas Development Institute, London.

Skutsch, M.M. (1983). *Why people don't plant trees: the socio-economic impacts of existing woodfuel programs*, Energy in Developing Countries Discussion Paper. Resources for the Future, Washington, D.C.

Skutsch, M.M. (1985). Forestry by the people for the people: some major problems in Tanzania's village afforestation programme. *International Tree Crops Journal*, **3**, 147–70.

Stolen, K.A. (1983). *Peasants and agricultural change in Northern Zambia*, Occasional Paper No. 4. International Development Program, Agricultural University of Norway, Ås.

TATF and ICRAF (1988a). *Agroforestry reseach project for the Sukuma agro-pastoral system in the unimodal upland plateau (Mwanza/Shinyanga region) of Tanzania*, AFRENA Paper No. 9. ICRAF, Nairobi.

TATF and ICRAF (1988b). *A blueprint for agroforestry research in the unimodal upland plateau of Tanzania*, AFRENA Paper No. 6. ICRAF, Nairobi.

Toit, R.F. du, Campbell, B.M., Haney, R.A. and Dore, D. (1984). *Wood usage and tree planting in Zimbabwe's Communal Lands*. Forestry Commission of Zimbabwe and the World Bank, Harare.

UNTF and ICRAF (1988a). *Agroforestry research project proposal for the Kigezi annual montane food crop system in the highlands of Uganda*, AFRENA Paper No. 11. ICRAF, Nairobi.

UNTF and ICRAF (1988b). *Agroforestry potentials for the land-use systems in the bimodal highlands of eastern Africa, Uganda*, AFRENA Paper No. 4. ICRAF, Nairobi.

Vedeld, T. (1983). *Social, economic, and ecological restraints on increased productivity among large circle chitemene cultivators in Zambia*, Occasional Paper No. 2. International Development Program, Agricultural University of Norway, Ås.

Vonk, R.B. (1983). *Report on a methodology and technology generative exercise, Kathama agroforestry project*. ICRAF and Wageningen Agricultural University, Nairobi and Wageningen.

Wachira, K.K. (ed.) (1987). *Women's use of off-farm and boundary lands*. ICRAF, Nairobi.

Wanjama, L.N., Buck, L.E. and Mueller, E.U. (1988). Case study of the PIASP Project Musgusa-Rwanda. Unpublished draft prepared for the Agroforestry Monitoring and Evaluation Methodology project. ICRAF, Nairobi.

Wanjohi, B. (1987). Women's groups, gathered plants and their agroforestry potentials in the Kathama area. In *Women's use of off-farm and boundary lands* (ed. K.K. Wachira), pp. 61–113. ICRAF, Nairobi.

Warner, K. (1993). *Patterns of farmer tree growing in eastern Africa: a socio-economic analysis*, Tropical Forestry Paper No. 27. Oxford Forestry Institute and ICRAF, Oxford and Nairobi.

Wilson, K.B. (1989). Trees in fields in southern Zimbabwe. *Journal of Southern African Studies*, **15**, 369–83.

Wood, P.J. (1966). *A guide to growing trees in Sukumaland*, Technical note (new series). Silviculture Section, Lushoto.

WRI/UNEP/UNDP (1990). *World Resources 1990–91*. Oxford University Press.

Young, A. (1989). *Agroforestry for Soil Conservation*. CAB International and ICRAF, Wallingford and Nairobi.

ZATF and ICRAF (1988). *Agroforestry research project for the maize-livestock system in the unimodal upland plateau (Chipata and Katete) in the Eastern province of Zambia*, AFRENA Paper No. 10. ICRAF, Nairobi.

ZATF and ICRAF (1989). *Agroforestry potential in the unimodal upland plateau of Zambia*, AFRENA Paper No. 7. ICRAF, Nairobi.

Part III
Factors influencing farmer decisions

6 Meeting household needs: farmer tree-growing strategies in western Kenya

Sara J. Scherr[1]

6.1 Introduction

A major justification for recent support of agroforestry and social forestry has been their potential contribution to rural welfare (Chapter 1, this volume; Gregersen *et al*. 1989; Falconer 1990). Such contributions include satisfaction of subsistence needs (for instance, food, fuel, building materials), substitution for purchased farm inputs (such as live fencing, animal fodder, green manure), opportunities to supplement cash income through sale of raw or processed tree products, and social uses (such as amenity plantings, shade, privacy, boundary markers). The potential for enhancing farm assets has also been highlighted, as trees and shrubs may be used to rehabilitate eroded or degraded soil and water resources, and timber trees may be used as stores of value for household savings, again without placing significant demand on households' scarce cash resources (Chambers and Leach 1989).

As a conceptual framework for designing interventions, however, the 'household needs' approach has had mixed success. Most smallholders in the tropics are deeply involved in the cash economy, and increasingly rely upon markets to supply household consumption goods and production inputs. A tree's potential to satisfy household needs for a particular product or service by no means assures that the farmer has any incentive to satisfy those needs through tree growing. Other sources of supply (gathering, purchase), product substitutes (kerosene for woodfuel, barbed wire for fencing), and income-earning opportunities (off-farm labour or new crops) may be more attractive options. There may also exist constraints to the adoption of agroforestry, even where incentives are attractive (Vosti *et al*. 1991).

This chapter draws from the theoretical and empirical insights of agricultural adoption studies to assess patterns of agroforestry adoption in meeting household needs. Its purpose is to illustrate the nature of adoption behaviour, and the research challenges, using a case study of agroforestry in Siaya and South Nyanza Districts of western Kenya. The first section poses some policy questions and summarizes the analytical framework which was used for the study. The second section presents results from the case study. The third section examines farmers' approaches to risk management in agroforestry adoption. The final section discusses the implications of the analysis for rural development and forestry policy, and for agroforestry project design.

6.2 Analysis of household tree-growing strategies

The discussion and design of agroforestry and social forestry interventions, by both government and non-governmental organizations, has generally been based on partial and static analysis of tree product use and demand by households. Scarcities are identified (such as, the 'fuelwood gap') and tree-growing interventions are proposed to address those scarcities. Typically, little attempt is made to examine the adaptations to scarcity already being made by farmers. The dynamics of farmers' economic response to scarcity and abundance and to supply and demand have been little recognized (Chapter 7, this volume). Yet this dynamic will determine the type of species, sites, intensity and systems which are likely to be adopted and should properly guide intervention efforts.

Indeed, one of the more important findings from recent research has been the remarkable degree to which farmers in many regions have intensified tree-growing activities in recent decades, without significant external intervention (Chapters 2, 3 and 9 in this volume; Bradley 1991). This evidence appears to refute the earlier view that agroforestry systems are a vestige of pre-capitalist rural economies destined to disappear with the rise of modern economies.

The analysis of farmer incentives for tree growing clearly requires a more sophisticated analytical framework. Two bodies of literature on land-use change are of particular value: 'livelihood strategy' and 'induced innovation'.

Livelihood strategy

Theories of 'livelihood strategy' are derived from earlier theories about 'household economy' and 'household decision-making' from economics (for instance, Baum and Schertz 1983; Singh *et al.* 1986) and from economic anthropology (Barlett 1980). Rather than assume that farmers are 'profit-maximizers', these theories focus on 'welfare maximization' and posit multiple household objectives, including secure provision of food and essential subsistence goods, cash for purchase of outside goods and services, savings (resources accumulated to meet future planned needs or emergencies), and social security (i.e., secure access to subsistence goods and productive resources). While seeking to meet these objectives, the farmer also seeks to reduce critical risk factors (Holden *et al.* 1991).

To achieve these aims, households select 'livelihood strategies' to pursue these objectives by use of the resources to which they have access. Resources include on-farm and off-farm lands, including perennial vegetation; household, hired, and other labour sources; and cash from product sales, off-farm labour, investments, remittances, etc. Both the resources available and the livelihood objectives change (often in predictable patterns) over the life cycle of the household. Despite high variability in the specific activities of different households, the basic strategies pursued by households with similar resources at similar stages in the life cycle tend to be similar. In investigating household response to subsistence, market or policy incentives for land and resource management, it is usually possible to identify groups of households with similar responses (Scherr 1985).

Tree-growing strategies in turn are determined by farmers' overall livelihood strategies and resource base (Rocheleau 1987). Farmers may be keen to grow timber trees for savings if they have no superior strategy for savings, while they reject growing trees for cash income if they already have a successful strategy for earning cash income from off-farm labour or crops. Farmers with many family members of working age may be little interested in farm fuelwood production, while those with little household labour may place a higher value on time saved from fuelwood collection.

Strategies would be expected to change with changes in relative value of inputs, alternative tree outputs, input or output substitutes, labour and land productivity, and household assets. Financial discount rates, as well as the farmers' implicit discount rates for different types of farm activities will affect decision-making, as will the degree of uncertainty of receiving benefits (goods or services) in the future.

Induced innovation

The second part of the analytical framework used here is from the theory of 'induced innovation' (Boserup 1965; Binswanger and Ruttan 1978; Ruthenberg 1980). While livelihood strategies are a response to socio-economic conditions at any point in time, 'induced innovation' theory focuses on historical changes in socio-economic conditions (particularly increased population density and market development) which lead to changes in relative input and output prices, and therefore the types of strategies and specific options considered by farmers.

Four types of long-term pressures may induce farmers to intensify tree husbandry. Declining access to wooded land due to extended settlement, loss of communal lands or restrictions on forest access may reduce tree product supply. Increasing demand for tree products may result from population growth, new tree uses or products, or demand from external markets. Increasing population density and declining farm size may create a social need for trees and shrubs as fences or boundary markers. Finally, land users may respond to declining land quality by protecting or planting trees and shrubs for windbreaks, soil fertility, erosion control, waterways protection, grazing land rehabilitation, or by substituting woody perennials (e.g., bananas in Uganda or tea in Kenya) for row crops on erodible soils (Scherr 1992a).

Intensification historically proceeds as a result of local innovation (both technical and institutional), as well as by extensively borrowing and adapting innovations from other areas. As the value of tree products or services rises, farmers are willing to consider increasingly costly means for acquiring them, but they will still tend to seek the least costly solution. This may involve tree planting, but may just as likely involve substitution of inferior tree products, or non-tree substitutes, or improved management of natural vegetation. In more extensive farming systems or activities, farmers will tend to adopt new practices only where expected returns are much higher than inputs, while in very intensive systems, farmers may value more marginal improvements, or higher input/output ratios.

Some policy questions

These dynamic factors determining farmer tree-growing practices suggest key policy questions, while complicating their analysis.

- Should agroforestry and social forestry programmes emphasize tree growing for cash or subsistence use? To what extent can we expect commoditization to determine future land use?
- How effective in improving family welfare, consumption or security are agroforestry/social forestry interventions relative to other types of policies?
- Should agroforestry and social forestry programmes concentrate on meeting specific household needs (e.g., food or fuel), or should they provide support for a wide range of uses?
- How can external development efforts complement and strengthen endogenous processes of agroforestry intensification?
- Should agroforestry or social forestry interventions be targeted to the poor? If so, how?
- In meeting household needs, how much emphasis should be placed on new tree planting and how much on management of and access to existing woody resources?
- What ancillary policies need to be implemented to increase the probability of success in agroforestry projects?

6.3 Historical changes in farmer tree growing in western Kenya

The rural economy of Siaya and South Nyanza

Siaya and South Nyanza Districts lie between Lake Victoria and the western highlands of Kenya. Moving from the lake, altitude increases from 1000 to 1500 m, and rainfall increases from 800 to 2000 mm per year, in a bimodal distribution (Fig. 6.1). These districts are populated mainly by the Luo people, a Nilo-Hamitic group which migrated from the north to Kenya during the 1500s and 1600s. Frontier settlement began in Siaya and continued into South Nyanza through the early 1900s. The economy was originally based on livestock herding and fishing, with subsistence agriculture based on millet. With permanent settlement in the 1800s, rising population densities, animal disease problems, and proximity to long-settled agricultural populations in the western highlands of Kenya, crop production became increasingly important (Scherr, 1993).

During the Colonial period, there was extensive land-clearing and settlement. The Luos learned to grow new crops, and the area became an important maize-exporting region. Use of ox-plough for cultivation predominated in the dry zones and hoe in the wetter zones. After Independence, as fallow periods shortened and land use intensification increased with little external input use, agriculture began a period of decline. Soil fertility and crop yields declined sharply. Fertilizer use was low due to high fertilizer–crop price ratios and poor yield response to fertilizer use. Average population density rose from very low levels in the 1800s to 100 people /km^2 by 1948 and to 250 people /km^2 by 1990 (with much higher densities of over 400 people /km^2 in the high rainfall zone; Scherr, 1993). Farm size in the higher potential zones declined through subdivision among sons, and there was large-scale labour migration to urban centres and more dynamic agricultural zones.

Historical changes in farmer tree growing

Three agroforestry periods can be identified. In the late 1800s and early 1900s, the farming system was oriented toward subsistence production. Farmers relied for fuelwood, building wood, medicines, food and fodder on tree products gathered or managed from naturally growing trees. With permanent settlement, farmers established farm and homestead boundary hedges of sisal (*Agave sisalensis*) and euphorbia (*Euphorbia terucalli*). Scattered fruit and timber species were protected in farm fields, and some especially valuable species were domesticated. Several exotic fruits, e.g., mango (*Mangifera indica*) and hedging species introduced by traders and travellers diffused widely and quickly. Colonial efforts to promote farm woodlots, however, were largely rejected, except for operators of a number of larger holdings.[2]

By the mid-1900s, deforestation and permanent settlement had sharply reduced off-farm land and tree resources. Increased population meant higher demand for wood products. Building-quality wood became scarce. In response, farmers began to domesticate local tree species and to adopt exotic timber species. Interstitial tree planting, i.e., planting in underused parts of the farm, such as around homesteads, or along field pathways or borders, became common. Few external inputs were used other than seedlings of a few, mainly exotic, species grown in central government nurseries. Naturally

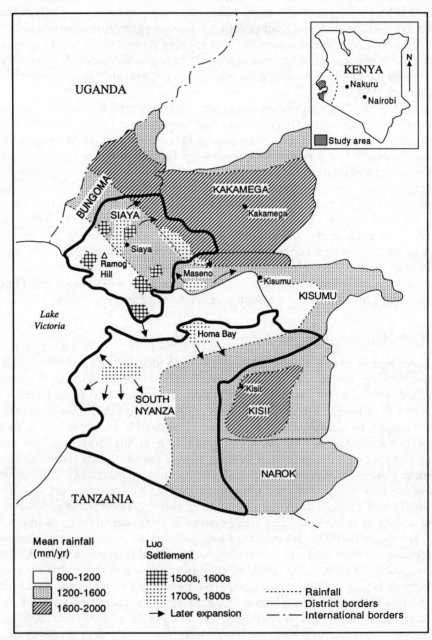

Fig. 6.1 Historical settlement and rainfall patterns in Siaya and South Nyanza Districts (from Scherr 1993).

growing trees were more intensively managed and timber trees were planted in non-cultivated areas of the farm, field borders, or occasionally woodlots. Farmer interviews and archival records[2] report that, to counter declining crop yields, some farmers began to enrich short fallows by scattering seed of indigenous nitrogen-fixing shrubs, and to use leaves from trees to mulch crops.

Agroforestry after Independence

Since Independence, agroforestry practices have intensified with the ongoing transition to permanent cropping, the disappearance of communal tree resources, and the rise of first, barter, and later, local cash, markets for woodfuel, poles, seedlings, and fruit. Patterns of intensification are illustrated by data from a study of agroforestry practices among Siaya and South Nyanza farmers participating in CARE International's Agroforestry Extension Project[3] (AEP) between 1985 and 1989 (Scherr and Alitsi 1991).

The objective of the AEP was to address household needs for food, fuelwood and construction wood, for which it offered an unusually broad 'menu' of agroforestry practices. Its philosophy was to base interventions on practices or species already familiar to farmers, to encourage incremental, adaptive change, to enhance the diversity of farmers' systems, and to intensify land use (Buck 1990). Their participatory approach placed most decision making about species, uses, planting sites and management with farmers. By 1990, the AEP had worked with 3000 farmers in 280 farmers' (mainly women's) groups, and 300 primary schools. Agroforestry practices on farms of a 14 per cent sample of project participants were surveyed.

These surveys showed that farmers were willing to intensify significantly their level of tree husbandry. Prior to the project, farms (which were typically 1–2 ha in size), had an average of 227 free-standing trees and 248 m of dense hedge.[4] During the 3-year average period of project participation, farmers increased the average number of free-standing farm trees by 125 per cent to 513, and the average length of hedge increased 29 per cent to 317 m. For farm plots with agroforestry, average tree density rose from 220 to 504 trees /ha, or by 129 per cent. Despite the very high coefficients of variation in the survey, these differences were significant at the 0.1 level (Tables 6.1 and 6.2).

Pressures for land use intensification are revealed by locations selected by farmers for tree growing (Table 6.3). Of all free-standing trees censused on farms, 25 per cent were found in the homestead, 39 per cent in or around cropland, and 17 per cent in or around pastures. Other farm niches conventionally considered most suitable for tree growing (paths, boundaries, woodlots and woody fallows) were only minor sites for tree growing. Dense hedges were found primarily around homesteads and cropland. Essentially, with declining farm size, many farmers had exhausted the potential for interstitial planting.

The cropland became the dominant site for tree planting. In individual and

Table 6.1 Average farm numbers and densities of free-standing trees

	Trees established before intervention	Trees established after intervention	Total	Significance[a]
Number of trees (trees per farm)				
Mean	227	286	513	*
Standard deviation	362	417	579	
Coefficient of variation	159	146	113	
Median	99	166	348	
Minimum	0	2	11	
Maximum	3 073	4 676	4 899	
Number of farms with trees	331	335	335	
Per cent of sample	98	100	100	
Tree density (trees per ha)[b]				
Mean	220	464	504	*
Standard deviation	816	2 336	2 209	
Coefficient of variation	371	504	438	
Median	110	153	257	
Minimum	0.3	0.1	0.8	
Maximum	14 500	40 000	40 000	
Number of farms with trees	330	333	335	
Per cent of sample	98	99	100	

Notes: * significant to 10 per cent. Totals are not the simple sum of before and after interventions. In some cases, farmers established trees both before and after interventions or in only one period. [a] Measured significant difference between pre-intervention numbers and total numbers of trees. [b] Densities calculated only for plots with some form of agroforestry practice. Because most plots supported agroforestry practices, estimates closely reflect densities of trees per total farm area.

focus group interviews, farmers suggested that this arose from three factors: a desire to better protect and manage more valuable trees, declining farm size which reduced the total area of non-cropped land, and increased use of tree species suitable for intercropping, such as *Leucaena* or *Sesbania*.

Table 6.3 also shows increases in tree densities. Most new trees were planted in fields where agroforestry was already being practised. Densities in homesteads and boundaries more than doubled, while densities in cropland tripled. Spatial configurations for tree growing reflected higher densities. Of all free-standing trees growing in 1989, only 27 per cent were in traditional scattered arrangements. Some 37 per cent were planted in lines, 21 per cent in small blocks, 9 per cent in linear intercropping (alley-cropping), and 6 per cent in systematic mixed intercropping.

More intensive methods of tree establishment were also being used. Of

Table 6.2 Average length and density of dense hedges ($n = 335$)

	Hedges established before intervention	Hedges established after intervention	Total[b]	Significance[a]
Hedge length (metres per farm)				
Mean	248	67	317	*
Standard deviation	380	179	414	
Coefficient of variation	153	265	131	
Minimum	0	0	0	
Maximum	4 170	1 800	4 170	
Median	127	0	204	
Number of farms with hedges	232	123	272	
Per cent of sample	69	39	81	
Hedge density (metres per ha)[a]				
Mean	369	178	461	*
Standard deviation	491	502	611	
Coefficient of variation	133	282	133	
Minimum	0	0	0	
Maximum	5 049	6 000	6 000	
Median	250	0	325	
Number of farms with hedges	232	123	272	
Per cent of sample	69	39	81	

Notes: * significant to 10 per cent. [a] Densities calculated only for plots with some form of agroforestry practice. Seven outliers (densities in excess of 10 000 square metres per ha) have been removed. [b] Totals are not the simple sum of before and after intervention. Different numbers of farmers established trees before and after intervention.

free-standing trees established prior to the AEP, only 30 per cent were from seedlings; another 9–10 per cent each were by vegetative propagation, direct seeding, and transplanted wildings, with a large share (42 per cent) by natural establishment. Dense hedges were mainly established vegetatively (42 per cent) or by natural establishment (30 per cent). During the project period, the proportion of free-standing trees established by seedlings shot up to 83 per cent, although this is a more labour-intensive method (Table 6.4).

6.4 Changing patterns of household tree use

Why, under conditions of increasing land scarcity, did Siaya and South Nyanza farmers increase the numbers and land area in trees? The qualitative and quantitative information of this study suggests that the predominant reason was to obtain critical consumption goods which would otherwise have to

Table 6.3 Tree density by type of land use

Tree density by land use, in trees per ha (per cent of all trees) ($n =$)	Before intervention (330)	Total (335)	Significance
Homestead (25% of total trees)			
Mean	209	478	**
Standard deviation	314	586	
Coefficient of variation	150	122	
Median	103	271	
n	292	311	
Cropland (39%)			
Mean	127	362	**
Standard deviation	191	524	
Coefficient of variation	151	145	
Median	64	213	
n	282	310	
Grazing land (17%)			
Mean	407	495	**
Standard deviation	733	723	
Coefficient of variation	180	146	
Median	146	225	
n	138	150	
Woodlots and woody fallows (16%)			
Mean	2916	4413	*
Standard deviation	4272	4875	
Coefficient of variation	146	110	
Median	1018	2433	
n	47	77	
Boundaries and paths (2%)			
Mean	407	871	
Standard deviation	616	1178	
Coefficient of variation	151	135	
Median	130	177	
n	11	20	
Other ($< 1\%$)			
Mean	323	1920	
Standard deviation	411	2088	
Coefficient of variation	127	137	
Median	323	743	
n	2	4	

Notes: ** significant to 5 per cent. * significant to 10 per cent. These figures do not include dense hedges. Outliers have been removed.

Table 6.4 Tree establishment methods

Establishment method, by planting configuration	Before intervention		After intervention		Total	
	No.	%	No.	%	No.	%
Free standing trees						
Direct seeding	7 904	10	6 862	7	14 767	9
Seedling	22 574	30	79 242	83	101 818	59
Cutting or sucker	7 194	9	1 952	2	9 146	5
Transplanted wilding	6 682	9	2 027	2	8 710	5
Naturally-growing	31 661	42	5 836	6	37 603	22
Unknown	49	<1	5	<1	54	<1
Total	76 064	100	95 924	100	172 098	100
Dense hedges						
Direct seeding	3 890	4	3 844	12	7 734	6
Seedling	12 589	13	12 202	38	24 961	19
Cutting or sucker	40 637	42	11 415	36	52 052	40
Transplanted wilding	11 238	12	1 217	4	12 455	10
Naturally-growing	28 886	30	3 066	10	31 952	25
Unknown	0	0	0	0	0	0
Total	97 240	100	31 744	100	129 154	100

Free-standing trees are measured in number of trees. Dense hedges are measured in metres. Reported numbers constitute 14 per cent of AEP participant farms.

be obtained by paying scarce cash. A second reason was to diversify their sources of cash income. A third reason was to protect food security in the face of declining crop yields.

Evidence of change in tree and species preferences

Let us look first at some of the evidence of shifting use preferences (Table 6.5). If we assess farm trees established prior to CARE intervention, we see that most free-standing trees were used primarily for building poles and fuelwood. Live fencing and boundary markers were also important uses. Fuelwood was a very common secondary use. About 23 per cent of trees were planted or protected for a wide range of uses: timber, fruit, shade, windbreaks, stakes, medicines. Of dense hedges, half were used primarily for live fencing, with fuelwood and boundary markers the main other uses, and 15 per cent of hedges for a variety of other uses. Animal fodder was rarely listed by farmers as a primary or secondary use of farm trees, despite the prevalence of goats. Off-farm pasture and fallow resources remained the primary feed sources for these animals. Browsing remains a problem for tree growing in the dry zones, where dry season grazing in cropland is prevalent.

Major changes can be observed for recently established trees. First is that the proportion of free-standing trees rose significantly compared with dense hedges. There was also a major decline in fuelwood as a principal use. This reflects a major shift from harvesting of fuelwood from naturally-growing trees, which had no other significant use, to harvesting as a secondary product from trees planted for another use. Despite the high coefficients of variation, this change was statistically significant at a 0.05 level (*t*-test) for both free-standing trees and dense hedges.

Building poles were by far the dominant use of trees established in recent years, together with timber accounting for nearly half of all trees. Also notable was the high proportion (15 per cent) of free-standing trees grown for green manure (mainly *Leucaena* alley-cropping), typically substituting for increasingly scarce manure on more marginal fields. There were functionally significant increases in numbers of trees grown singly or in small plots for fruit, shade, medicine, timber, and stakes. The increase in trees primarily grown for animal fodder, particularly from dense hedges, reflects emerging interest in dairy production in the high rainfall zone, and dry season fodder reserves in the dry zone.

These revealed farmer preferences are confirmed by their responses to a question asking them to rank five species and use preferences for future planting. Households wanted at least one species for building poles (72 per cent of respondents), fruit (62 per cent), green manure (44 per cent), shade (30 per cent), fencing (19 per cent), fuelwood (16 per cent), timber (16 per cent), ornamental (12 per cent), fodder (7 per cent), and other (7 per cent). Including all five preferences per respondent, most requests were for building poles (22 per cent), fruit (21 per cent), timber (13 per cent), green manure (11 per cent), shade (9 per cent), fuelwood (5 per cent), fencing (4 per cent), other (9 per cent), and not specified (6 per cent).

Changes in tree use preferences were reflected in significant shifts in tree species choice (Tables 6.6 and 6.7). Species choice was further influenced by improved availability of seed and nursery methods for seedling production (even of indigenous species which were not previously planted) induced by the project. Several features are worth highlighting. First is the very high number of total species which are commonly found, although there was a significant increase in the dominance of the five principal species. Table 6.8 shows that overall, while five species account for half of free-growing trees, 25 account for another 43 per cent, and 137 other species account for the remaining 5 per cent.

For several new exotic species, proportions more than doubled: *Cassia siamea*, *Leucaena leucocephala*, *Melia azadirach*, *Thevetia peruviana*, *Terminalia mentalis*, citrus, papaya, and jacaranda. There were significant reductions in the proportion of several important indigenous or traditional species: sisal, euphorbia, lantana, *Albizia coriaria*, *Terminalia brownii*, and *Combretum molle*. Overall, however, farmers seem to have chosen a practical

Table 6.5 Changes in primary uses of trees on farms

Planting configuration and uses	Before intervention No. (n = 330)	% 98	After intervention No. (n = 333)	% 99	Total No. (n = 335)	% 100
Free-standing trees						
Ornamental	116	<1	433	<1	549	<1
Boundary marker	3 875	5	323	<1	4 198	2
Building poles	26 637	35	36 348	38	62 985	37
Charcoal	323	<1	32	<1	355	<1
Animal fodder	681	1	1 037	1	1 718	1
Live fencing[a]	7 334	10	8 193	9	15 527	9
Fruit	5 091	7	7 441	8	12 635	7
Fuelwood	18 352	24	9 948	10	28 306	16
Green manure	288	<1	13 967	15	14 255	8
Rope	983	1	456	<1	1 439	<1
Soil conservation	254	<1	1 590	<1	1 844	<1
Seed	1	<1	15	<1	16	<1
Shade	3 039	4	3 733	4	6 773	4
Medicine	656	1	1 030	1	1 686	1
Stakes or fitos	627	1	1 027	1	1 654	1
Timber	6 443	8	9 011	9	15 454	9
Windbreaks	1 264	2	1 259	1	2 523	1
Total	75 964	100	95 843	100	171 917	100
Dense hedges						
Ornamental	10	<1	10	<1	20	<1
Boundary marker	14 031	14	1 640	5	15 671	12
Building poles	3 754	4	2 297	7	6 051	5
Charcoal	260	<1	260	<1	520	<1
Animal fodder	192	<1	1 205	4	1 397	<1
Live fencing[a]	50 929	53	22 951	72	73 880	57
Fruit	1 316	1	466	1	1 782	1
Fuelwood	17 796	18	1 229	4	19 025	15
Green manure	92	<1	687	2	779	1
Rope	3 536	4	30	<1	3 566	3
Soil conservation	75	<1	60	<1	135	<1
Shade	140	<1	410	1	550	<1
Medicine	947	1	181	1	1 128	1
Stakes or fitos	2 325	2	330	1	2 655	2
Timber	52	<1	93	<1	145	<1
Windbreaks	1 384	1	120	<1	1 674	1
Total	96 839	100	31 969	100	128 978	100

Notes: [a] Most live fences are established as dense hedges. Their length in metres was measured, rather than counting individual trees. Figures were derived from a 14 per cent sample.

Table 6.6 Changes in free-standing tree species grown by farmers

Species and uses	Before intervention		After intervention		Total	
	No.	%	No.	%	No.	%
(n =)	(330)	98	(333)	99	(335)	100
Main forestry trees						
● *Eucalyptus* sp.	12 698	17	20 771	22	33 469	19
● *Cassia siamea*	2 282	3	5 798	6	8 707	5
Sub total	14 980	20	26 569	28	42 176	25
Main agroforestry trees						
▲ *Markhamia lutea*	10 747	14	12 417	13	23 164	13
◆ *Leucaena leucocephala*	348	0	16 638	17	16 986	10
Subtotal	11 095	15	29 055	30	40 150	23
Other trees for wood and leaf						
Multiple naturally growing species[a]	6 376	8	1 600	2	7 976	5
▲ *Agave sisalensis*	6 096	8	1 760	2	7 856	5
● *Cupressus lusitanica*	2 178	3	5 270	5	7 448	4
▲ *Euphorbia tirucalli*	5 065	7	1 676	2	6 741	4
▲ *Lantana camara*	5 252	7	636	1	5 888	3
◆ *Terminalia peruviana*	2 915	4	2 731	3	5 844	3
● *Cassia spectabilis*	1 141	2	3 031	3	4 558	3
● *Grevillea robusta*	230	<1	2 867	3	3 097	2
▲ *Acacia* sp.	2 843	4	90	<1	2 936	2
▲ *Albizia coriaria*	1 711	2	413	<1	2 124	1
◆ *Melia azadirach*	531	1	1 454	2	1 985	1
◆ *Parkinsonia aculeata*	7	<1	1 849	2	1 856	1
▲ *Sesbania sesban*	374	<1	1 381	1	1 755	1
◆ *Callitris robusta*	335	<1	1 133	1	1 468	1
◆ *Calliandra calothyrsus*	7	<1	1 356	1	1 363	1
▲ *Euphorbia candelabrum*	646	1	452	<1	1 098	1
▲ *Sesbania bispinosa*	0	0	785	1	785	<1
▲ *Bridelia micrantha*	424	1	349	<1	773	<1
▲ *Terminalia brownii*	661	1	162	<1	773	<1
▲ *Combretum molle*	731	1	38	<1	769	<1
◆ *Croton megalocarpus*	373	<1	249	<1	622	<1
Subtotal	37 896	50	29 282	31	67 715	39
Fruit trees						
▲ *Psidium guajava*	2 659	3	1 505	2	4 265	2
● *Citrus* sp.	670	1	2 461	3	3 131	2
● *Carica papaya*	402	1	2 001	2	2 403	1
▲ *Mangifera indica*	1 043	1	690	1	1 734	1
Subtotal	4 774	6	6 657	7	11 533	7

Table 6.6 (*Cont.*)

Species and uses	Before intervention		After intervention		Total	
	No.	%	No.	%	No.	%
Ornamental/shade trees						
• *Jacaranda mimosifolia*	202	<1	887	1	1 089	1
♦ *Terminalia mentali*	50	<1	626	1	676	<1
Subtotal	252	<1	1 513	2	1 765	1
Minor species	7 067	9	2 847	3	8 749	5
All free-standing species	76 064	100	95 923	100	172 088	100

Notes: ▲ Species indigenous or introduced to Siaya and South Nyanza Districts prior to 1900. • Species introduced to Siaya and South Nyanza Districts between 1900 and 1950. ♦ Species introduced to Siaya and South Nyanza Districts between 1971 and 1989. ª Multiple naturally growing trees consist of agroforestry practices in a highly variable species mix. Most were found in bush fallows or in dense hedges around homesteads, and received little management. Numbers were estimated rather than counted

mix of indigenous, colonial and recently introduced species, which reflect their use preferences. If we cluster species by their predominant use, we find the biggest change was an increase in species considered by farmers to be most suitable for intensive intercropping: *Markhamia* and *Leucaena*.

Trees to secure 'forest' products

The survey results, together with farmer interviews, indicate that most farm trees in Siaya and South Nyanza were planted or protected for their contribution to direct household use for consumption, amenity or social value, or farm inputs. Farmers' willingness to invest greater household resources in tree growing is primarily a response to increasing scarcity of these products/inputs in naturally-growing vegetation on- and off-farm relative to the needs of a growing population, and relative to available substitutes. At the same time, extension assistance helped to reduce some of the costs and constraints of on-farm tree planting.

Table 6.9 shows, for example, that although 66 per cent of households obtained some fuelwood supplies by collecting from off-farm resources, for only 20 per cent of households was this the primary source. Naturally-growing trees on farms were universally used as a fuelwood source, and were the primary source for 41 per cent of households. But planted trees on farms were overtaking naturally-growing trees as the primary source of fuelwood. Similarly, although most households still used some poles naturally-growing off-farm (32 per cent) or on-farm (76 per cent), these two sources now were the primary source for only a minority of project-assisted farmers. A majority of households (56 per cent) relied primarily on planted farm trees.

Household response to scarcity may take many forms, only one of which is to intensify tree production. Other options were to meet new demands through inter-regional trade, to substitute non-tree products for the same uses, to increase use efficiency, and/or to reduce consumption levels.

But the critical shortage of household cash income (both farm and non-farm), together with high transport costs of moving goods from other areas of Kenya to this region, effectively limited the potential for Luos to meet household needs by importing tree products from outside the region or by finding non-farm substitutes (e.g., kerosene for fuelwood).

The third option, to increase the efficiency of use of tree products, was practised to some extent. There were changes in house construction designs to reflect changing timber supplies. Community standards for the

Table 6.7 Changes in hedging species grown by farmers

	Before intervention		After intervention		Total	
	(m)	%	(m)	%	(m)	%
(*n* =)	(241)	85	(133)	45	(284)	100
Main hedging species						
▲ *Lantana camara*	35 992	37	10 185	32	46 177	36
▲ *Euphorbia tirucalli*	29 517	30	10 826	34	40 343	31
▲ *Agave sisalensis*	12 367	13	2 040	6	14 407	11
◆ *Thevetia peruviana*	3 417	4	4 350	14	7 937	6
Multiple naturally-growing species	6 805	7	0	0	6 805	5
● *Cupressus lusitanica*	1 958	2	1 296	4	3 254	3
▲ *Draecaena steudneri*	1 190	1	0	0	1 190	1
▲ *Annona senegalensis*	816	1	295	1	1 111	1
▲ *Harrisonia abyssinica*	1 025	1	2	<1	1 027	1
◆ *Leucaena leucocephala*	300	<1	698	2	998	1
Main species of dense hedge	93 387	96	29 692	94	123 249	95
Minor species of dense hedge	3 823	4	2 052	6	5 875	5
Total length of dense hedge on 335 farms	97 210	100	31 744	100	129 124	100

Notes: ▲ Species indigenous or introduced to Siaya and South Nyanza Districts prior to 1900. ● Species introduced to Siaya and South Nyanza Districts between 1900 and 1950. ◆ Species introduced to Siaya and South Nyanza Districts between 1971 and 1989.

level of household fuelwood stores reportedly declined. Efforts were made by CARE and others to introduce more efficient cookstoves (Vonk 1986), although farmer interviews indicated that these had little impact on overall fuelwood use.

The fourth alternative of reducing average levels of household consumption of tree products was most widely practised (Muturi 1991). Farmers reported in interviews the increasing use of 'inferior' fuelwood species, lower-quality timber in house-building, declining numbers of household animals (requiring fewer fodder resources), and declining use of bush fallow and animal manure which resulted in declining crop yields. These practices were clearly associated with a decline in living standards, which could be countered by most farmers, under the economic conditions prevailing in Siaya and South Nyanza, only by increasing local tree supplies.

Trees for food security

Farmers increasingly used trees to enhance food security. There were modest, but functionally important, increases in tree growing for fruit. The wide variety of indigenous and exotic fruits grown (27 species reported in the survey), in small numbers per farm, suggest their probable importance in household consumption. Fruits are particularly used as a food for children, and to tide over during the pre-harvest period.

In terms of total woody vegetation, however, the expanded role of trees as an input or service for food production was more notable. Two key

Table 6.8 Tree species diversity on farms, 1989

Species	Per cent of all free-standing trees
5 dominant species[a]	52
25 other common species	43
18 planted species with at least 50 trees counted[b]	2
51 planted species with fewer than 50 trees planted[b]	<1
68 species nearly always naturally growing[c]	3
167 total species	100

Notes: [a] *Leucaena leucocephala, Markhamia lutea, Eucalyptus, Agave sisalensis,* and *Cassia siamea.* [b] These include species for which at least 30 per cent of the individuals counted were planted, rather than naturally-growing. [c] Species for which more than 70 per cent of the individuals counted are naturally-growing. Excludes any species which was counted as planted, e.g., a species that was 50 per cent planted and 50 per cent naturally-growing.

Table 6.9 Use of planted trees for fuelwood and building poles, 1989

(*n* = 335)	Per cent of households	
Sources of household fuelwood supplies during past year	Sources used	Primary source
Naturally-growing trees on farm	94	41
Planted trees on farm	89	36
Collected off-farm	66	20
Purchased	36	3
Sources of household building poles during the past year		
Naturally-growing trees on farm	76	19
Planted trees on farm	85	56
Collected off-farm	32	4
Purchased	56	21

innovations were windbreaks and alley-cropping. A large proportion (15 per cent) of new trees were grown for green manure (mainly *Leucaena* alley-cropping and improved fallows of *Sesbania* sp., *Leucaena* and other nitrogen-fixing species). Farmer interviews suggest that these substituted for increasingly scarce animal manure on more marginal fields, or were used to rehabilitate highly degraded sites.

While alley-cropping is still treated by most farmers as an experimental technology (average area under alley-cropping in 1989 was only 500 m2), the fact that 44 per cent of CARE-assisted households were willing to test the system indicates both the desperate situation of yield decline and interest in innovation. Their use of alley-cropping for soil fertility management compares with 81 per cent reporting use of animal manure, composting (77 per cent), crop rotations (60 per cent), natural fallow (55 per cent), herbaceous fallows (18 per cent), and fertilizer (16 per cent). Alley-cropping was also used by 30 per cent of farmers as a soil conservation measure, although physical structures (53 per cent), grass strips (43 per cent) and trash lines (42 per cent) remained much more important.

While farmers reported that only 2 per cent of farm trees were grown primarily as a field windbreak (typically single-row), another 4 per cent performed a secondary function as a windbreak. In addition, many of the lines of trees around crop fields and homesteads, whose primary and secondary uses were for poles, boundary markers, fuelwood and other uses, also performed (and were designed) as field windbreaks. This was particularly important in the drier parts of the zone where wind damage to crops and desiccation can significantly reduce crop yields.

Tree products for the market

The apparent importance of household consumption as the initial driving force behind agroforestry intensification did not preclude farmers' interest in market opportunities. This is reflected most directly in farmers' growing of *Eucalyptus* sp. for urban construction pole markets and citrus and papaya for local and regional fresh fruit markets. Together, these primarily commercial species accounted for a quarter of all free-standing trees grown on farms in 1989.

Furthermore, inter-household trade in building poles has developed at the community level. Since at least the 1950s, there has been a barter economy in the more populated parts of Siaya and South Nyanza. With the monetization of the local economy through cash cropping and remittances from migrant labour, this has been transformed into local cash markets, whose importance continues to grown. Species other than eucalyptus, such as *Cassia siamea*, *Markhamia lutea*, *Cupressus lusitanica*, *Cassia spectabilis*, *Grevillea robusta*, *Thevetia peruviana* and *Terminalia brownii* were marketed locally. Some 56 per cent of the sampled farmers reported purchasing at least some of their poles for the last farm building constructed, while 21 per cent had used primarily purchased poles.

Markets have, for a long time, provided woodfuel supplies for small-scale industries, as well as charcoal for domestic consumption. Historically, mining companies and small-scale fish-drying enterprises near Lake Victoria were important consumers.[2] Currently, important purchasers also include schools, local sugar factories, brick-makers and illegal beer brewers.

Markets for household fuelwood have grown up more recently, apparently as a response to scarcity of natural sources, preference for higher quality supplies, rising cash incomes among better-off rural dwellers, and/or rising scarcity of women's time, especially those without resident husbands. Among the farmers surveyed, 36 per cent had purchased some fuelwood for household use in the previous year. For 3 per cent of households purchased fuelwood was the primary source.

Thus, while most farmers interviewed initially planted trees for building poles to reduce their cash expenditures for house construction, the marketability of any surplus supply was soon recognized. With average monthly incomes in rural Siaya households estimated in 1990 to be KSh 500 (at 30 Kenya shillings per US dollar), the periodic sale of 4-m building poles for KSh 10–30 represented significant income. Farmers also began to recognize the growing cash demand for other tree products among their neighbours and in local markets.

A 1987 survey of local market centres in Siaya showed many products being sold (Oduol and Akunda 1987), but sales within communities or between neighbours still predominate. Interviews with groups of farmers indicated that, by the early 1990s, the growing of poles, and to a lesser extent, other tree products, had become an important part of the overall cash strategies of many households.

The household survey indicated that 51 per cent of households had sold some building poles during the previous year, the highest proportion (60 per cent) in South Nyanza. Moreover, many households (47 per cent) had sold fruit, fuelwood (32 per cent), tree seedlings (22 per cent), tree herbs (10 per cent), tree seed (4 per cent) and other products (3 per cent). The extent of tree product sales was especially high in the low rainfall zones for fuelwood (43 per cent) and fruit (56 per cent), and higher for seedlings in the higher rainfall zone (24 per cent) and in South Nyanza (34 per cent).

There is evidence of high discount rates, at least for subsistence income. Extensionists have had to invest considerable effort in some areas to encourage farmers' groups to plant on their own farms the seedlings produced in group nurseries. Women's need for immediate cash income apparently often outweighed the higher expected future income which could be earned by planting the trees on the farm or in a group plot. This trade-off may be exacerbated by women's inability to control cash income generated from farm trees which, by Luo customary rules, are men's property.

It can be speculated that interest in agroforestry was high in this area in part because of the lack of alternative sources of cash income. Where pole and timber trees did have some suppressive effect on associated crops, the value of the lost crop (given low prices and low base yields) was low relative to the value of the tree.

Farmers' interest in trees for increasing crop yields (alley-cropping, improved fallows) may reflect a desire to reduce cash expenditures on food, rather than expectations that yields would rise high enough to represent significant sources of net cash income from crops. In the new cash cropping of robusta coffee in the higher rainfall areas, farmer interviews suggested that *Leucaena* and other nitrogen-fixing trees were being used to substitute for expensive fertilizers.

Variations in tree use

The summary tables and discussion above suggest trends in agroforestry and tree use. It must be emphasized, however, that there is considerable farmer-to-farmer variation in the species, configurations and uses. While all project-assisted nurseries together worked with a total of 46 tree species, each group typically produced only five to 10 species selected to meet the group's priority needs. (Of course, other sources of planting materials were also used.) Major factors influencing agroforestry practices included agroecological zone, farm size, wealth category, gender and household status of the group member, and stage of the household demographic cycle.

Agroecological zone affected agroforestry preferences in several ways. Climatic conditions influenced species selection, as different types of trees grew well in the wet and dry zones. For example, *Grevillea robusta* was found mainly in the wettest areas and *Acacia* sp. mainly in the dry. Moisture availability also affected the nature of tree-crop competition in intercrops. *Leucaena* intercrops were much more problematic in the dry zone, where *Leucaena* was grown more commonly for poles rather than green manure. Windbreaks were more common in the dry zone where most of the problems with desiccating and destructive winds were found. Greater problems of damage to trees from livestock in the dry zone influenced farmers' choice of planting sites. Survey data show a higher proportion of trees grown principally for fuelwood in the low rainfall zone, for fencing in the high rainfall zone, and for green manure in the medium-rainfall zone; only the fuelwood difference was statistically significant, at the 0.05 level. Alley-cropping was adopted by significantly more farmers in the medium rainfall zone for both soil conservation and fertility.

Because higher rainfall zones were also those with highest population densities, local tree product markets were most developed in these areas (Muturi 1991). At the same time, a much higher *proportion* of farmers in the low-rainfall zone reported cash sales of both fuelwood and fruit (statistically significant at the 0.006 and 0.02 levels).

Farm size also influenced agroforestry preferences. Where farms were larger, more trees could be incorporated into the farm away from the cropland, in un- or underexploited niches. This favoured the use of trees such as eucalyptus which are fast-growing, but led to serious yield declines when planted in or adjacent to cropland. On smaller farms, there was more tree-planting in homesteads and cropland, and greater interest in soil fertility-enhancing practices.

Overall *household wealth* was also associated with some differences in agroforestry use preferences, as shown in Table 6.10. Fuelwood is a much more common use among those considered 'poor' (*t*-test significant at 0.05 level). Those considered 'average' had a higher proportion of trees for fruit and timber (significant at 0.05 level between poor and average groups). Farmers considered to be significantly wealthier than others in the community invested more heavily in fencing (significant at 0.10 level). Alley-cropping was more commonly used as a soil fertility strategy by households of 'average' wealth.

There is also a notable inverse relationship between wealth rank and tree densities. Poor farmers have the highest densities with 446 trees per ha; average farmers have 347 per ha and wealthy farmers have 283. Also, approximately 90 per cent of the trees found on a per hectare basis on poor and average income farms were planted, compared with only 68 per cent for wealthy farms. Both of these findings are likely to reflect differences in access to farmland, with wealthy farmers having more land in fallow.

Household status and gender also affected agroforestry patterns. Of group members surveyed, a third overall and nearly half in the dry zone were men.

Table 6.10 Primary use of free-standing trees by household wealth class

				Household wealth class				
	Poor		Average		Wealthy		Total	
Use (n = 335)	Number (98)	Per cent 29	Number (224)	Per cent 67	Number (10)	Per cent 3	Number (335)	Per cent
Fuelwood	9 736	23	17 363	14	1 203	19	28 302	16
Fruit	2 197	5	9 829	8	573	9	12 599	7
Fencing	3 294	8	10 745	9	1 148	18	15 187	9
Building poles	16 313	38	44 920	37	1 634	26	62 867	37
Timber	1 926	4	13 085	11	414	7	15 425	9
Green manure	3 856	9	9 916	8	483	8	14 255	8
Subtotal	37 322	86	105 858	87	5 455	86	148 635	87
Other	5 900	14	16 114	13	905	14	22 919	13
Total	43 222	100	121 972	100	6 360	100	171 554	100

Notes: These figures do not include dense hedges. Totals are larger than the sum of the specific uses because some uses have been excluded. The total refers to all trees on farms. Figures are derived from a 14 per cent sample.

Another 41 per cent of group members were women whose husbands were resident on the farm. Thirteen per cent (more in the high rainfall zone) were women whose husbands worked away from home, and 14 per cent were women without husbands (Table 6.11).

The average number of trees on farms was significantly higher among households which were male headed households. These men have both the interest and authority to plant trees, and operate a larger farming area. Women without husbands had the fewest trees. Women with husbands away from home had more trees than women with husbands at home, and also showed the highest increase in tree numbers and densities during the period of extension assistance. It is possible that these women had greater autonomy for farm management with husbands away, or greater need for the supplemental income provided by trees. Women heads of households also tended to plant more trees in the homestead and cropland, and to plant fewer trees for timber and building poles (Bonnard and Scherr 1994).

The stage in the *demographic cycle* of the household also appears to have affected choice of agroforestry practices, although only qualitative information about this factor is available from interviews with farmers and extensionists.

Table 6.11 Influence of household status and gender on tree growing

Number of trees and density of trees found on farms	Gender and status of head of household				
	Male	Female, husband on farm	Female, husband away	Female, no husband	Overall
Number	105	137	44	47	(335)
Per cent of sample	31	42	13	14	(100)
Number of free-standing trees, per farm					
Mean	642	426	458	406	510
Standard deviation	597	471	557	392	574
Coefficient of variation	93	110	122	96	113
Minimum	69	13	11	28	11
Maximum	4150	2680	2576	1807	4150
Tree density (trees/ha)					
Mean	444	340	398	291	374
Standard deviation	444	285	341	340	442
Coefficient of variation	100	84	86	117	114
Minimum	8	21	6	23	6
Maximum	2571	1325	1561	1939	2571

Notes: *t*-tests indicate that there is a significant difference, to 10 per cent, in average number of trees between males and women with husbands on the farm and between males and women without husbands. The difference in average densities between males and women without husbands is also significant to 10 per cent.

There was little tree planting by young families, as in most cases a Luo son does not receive title to the land (and thus formal rights for him or his wives to plant trees) until his father dies. Families with many children still at home placed emphasis on food security, as well as regular cash income to pay school fees. With more abundant labour resources, they can adopt more labour-intensive agroforestry practices. Older farmers, meanwhile, whose children are grown, more often opt for lower-labour-demanding activities such as woodlots and border plantings for poles or homestead plantings.

6.5 Reducing the risks of agroforestry innovation

Agroforestry adoption can be associated with some important risk factors. These include potential loss of trees to drought, browsing or disease; suppressed production of associated crops, trees providing habitat for pests or diseases of crops, failure to achieve expected levels of production or protection, high potential costs of tree removal should the farmer choose not to use the agroforestry practice, future loss of rights to harvest tree products, possible theft of tree products, and possible conflicts with neighbours over border trees.

In considering strategies for using trees to address household subsistence needs among poor farmers, it is important to recognize that their adoption process for new species and practices will reflect a concern to carefully manage risk. The experience in Siaya and South Nyanza is instructive. Interviews with farmers, groups and extensionists indicated a number of mechanisms used to reduce risk.

One mechanism was to reduce the number of unknowns by integrating new components, uses or management practices into existing practices. For example, the locally-popular *Markhamia* was often used to test tree planting in lines and alley-cropping before new species were tested in those unfamiliar configurations. New nursery practices introduced with exotic species were often adapted by farmers for local species.

A second mechanism was to adopt new practices experimentally, with small numbers of trees or small areas under new agroforestry systems. For example, the 1988 farm survey showed that the average initial size of alley-cropping plots in Siaya was only 500 m². This small-sized plot could not be expected to have a significant impact on household food yields, but was appropriate for early learning, testing and evaluation by farmers of what was a fairly radical land use innovation. Other new tree species were tested in small areas as well, and farmers were, in fact, enthusiastic to learn about and test more species. In the case of long-lived trees, the act of tree establishment and even ongoing management and use cannot in themselves be taken as evidence of 'adoption,' but only of willingness to test the system.

A third mechanism for risk management was diversity. Farmers were interested in having several different tree species, even for the same use. In many cases, different species were in fact required for different niches or site conditions (e.g., *Eucalyptus* was preferred for commercial pole production away from crops and *Leucaena* was preferred for trees grown in crop fields). The survey documented a total of 173 different tree species on farms (see Table 6.8). The 30 most commonly planted accounted for 95 per cent of farm trees. But there were also 68 indigenous species nearly always naturally-growing, 18 planted species which were less common, and 51 rare planted species. Farmers both retained the most valued traditional species and actively adopted new species which met their needs.

Fourth, while farmers tended to select tree species for particular uses, they appreciated multi-purpose trees with multiple management options. These offered them the opportunity to take advantage of changing household needs and to offset their vulnerability to changing market prices. *Leucaena* was especially appreciated as different management practices could be used to produce predominantly green manure, fuelwood, building poles or fodder.

Management could respond flexibly to climatic conditions, for example, by diverting leafy biomass from crops to animals in seasons of low feed availability. Similarly, pole species which had significant side branching were not necessarily devalued for that reason, as the side-prunings were usable as fuelwood.

Fifth, farmers reduced risks associated with delayed returns from longer-rotation tree production (and accommodated higher discount rates) by incorporating components with early returns. For example, though farmers recognized the potentially higher value per unit volume of large-diameter (long rotation) timber, they opted increasingly for shorter rotation systems to provide early cash returns from poles. This objective also led to high-density tree planting (in both lines and blocks) so thinnings could be used for fuelwood, stakes or small-diameter poles during the initial 2–5 years after establishment. Another example is the intercropping of passion fruit vines or papaya (as well as annual crops) with pole trees to provide early income off the land.

Sixth, farmers were reluctant to invest scarce resources in agroforestry practices purely for environmental rehabilitation. Rehabilitation investments were implemented mainly where land degradation threatened the imminent loss of productive resources (severe erosion resulting in loss of land, or severe nutrient depletion in cropland resulting in very low yields).

Since household resources were insufficient to allow optimal management of all farm lands, farmers often wished to concentrate their resources on the best lands or best food-producing/income-earning options. The characteristics of agroforestry-based rehabilitation efforts, particularly their dual role in providing subsistence or marketable products, led to adoption where purely rehabilitative technologies might have been rejected.

6.6 Summary and policy implications

What lessons can we draw from the western Kenya experience in answering the policy questions posed earlier? To what extent should agroforestry and social forestry development efforts emphasize household subsistence needs? How can such efforts be most effective in meeting priority needs for poor households?

Selecting agroforestry objectives: subsistence goods, commodities or resource protection?

Current debates about agroforestry policy and strategy can be attributed in part to the highly divergent perspectives which led to original interest in the field (Steppler and Nair 1987). Some groups were concerned with rural welfare, and particularly what they saw as the bypassing of ecologically marginal regions and poorer farmers by the Green Revolution. They saw agroforestry as part of an alternative development strategy which would meet basic subsistence needs while reducing the vulnerability of the poor to market fluctuations and other risks.

Agroforesters coming out of the public and commercial forestry sectors emphasized the promising role of farm tree production in supplying regional and national wood product markets. Together with the already strong influence of those promoting commercial tree crops (tea, coffee, tannin, rubber, fruits), they focused on the market potential of trees. Yet a third group was drawn to agroforestry primarily as an approach to land use in ecologically 'fragile' areas, which would permit sustainable farm production while halting or reversing associated erosion and preserving tree cover.

All three perspectives, basic needs, and needs for cash crops and for environmental protection, are legitimate. However, to adopt any one as the sole foundation for agroforestry policy will almost certainly be a mistake. Different farming systems, different groups of farmers within a given system, and different trees on a given farm can benefit from different strategies.

Reflecting local development dynamics

Decisions about the strategies of agroforestry development programmes should instead be grounded in local development dynamics: the overall dynamics of socio-economic development, and the dynamics of agroforestry intensification. If scarcity of tree products has been documented, what are people's responses to scarcity? What are the implications of scarcity for household welfare? To what extent have responses included greater investment in tree husbandry? Are there alternatives to tree growing to alleviate the problem? How do the costs and benefits of tree growing compare with those alternatives? How do price, land use and other policies affect farm decisions?

To answer these questions requires some understanding of long-term trends in population density, general market integration, and land use. This further requires an accurate assessment of existing tree resources, ongoing patterns of change in regional household tree use and production, socio-economic and agronomic factors influencing these changes, and an estimate of likely future external demand for tree products. In Siaya and South Nyanza, agroforestry intensification arose in response to population and internal market pressures, while farmers' selected practices reflect factor availability and market access.

Agroforestry programmes have tended to focus on provision of planting materials. An assessment of regional opportunities and constraints might suggest instead efforts to improve marketing services, remove regulatory barriers to tree growing, organize natural woodlands management by community groups, remove barriers to access to existing tree resources, or link with other development projects. In western Kenya, there was greater scope for increasing farm than community tree resources. Farmers were already actively protecting valuable trees and needed help to decrease the costs of domestication and intensive planting. While they initially focused on self-sufficiency objectives, interest quickly turned to commercial opportunities, with consequent demand for greater assistance with marketing.

The Kenya data also suggest that we can expect a high degree of inter- and intra-household variability in the demand for specific tree products and services. These will be influenced by differences in resource access, opportunity costs, and varying strategies for achieving short- and long-term food and income security. Development programmes can address these broad needs, by supporting a 'basket of options', even where there is greater intensity of effort for particular purposes.

Resource conservation will continue to be a major justification for promotion and adoption of agroforestry. But evidence in Siaya and South Nyanza suggests that adoption is most likely where the conservation function is embedded in technologies which also, or even primarily, provide clear benefits to the household as product, service or source of cash. Over time, agroforestry intensification may be associated with significant reorganization of the landscape, reflecting changes in land managers' valuation of different land uses.

Targeting the poor

The western Kenya data demonstrate that agroforestry can be used by farmers at any income level. This land use approach can have particular value for poorer households, because of the potential for substituting for scarce land, labour and cash resources. Poorer households can be reached by directing extension and support efforts to promotion of trees to meet household needs, both for subsistence uses and substitutes for purchased inputs. More financially secure households may also be interested in some of these practices, but will be less likely to divert resources to subsistence production when they have an acceptable option to purchase.

More labour-intensive practices may be biased in favour of poorer smallholders. Interventions increasing family labour productivity at low capital cost benefit most poor households. Working directly with poor farmers' or women's groups (as in the AEP) can serve to direct interventions to their needs.

General programmes to encourage cash tree crops could easily bypass the poor, if poorly designed. But cash tree crops can benefit the poor (who usually have critical cash constraints) so long as certain conditions hold

or can be developed. These include low costs of entry, a short waiting
period until cash returns begin, and market channels which serve small-
scale, as well as large-scale producers. Risks associated with the practice should be low, or serve to reduce other farm risks. Poor farmers
benefit from the opportunity to establish production incrementally. Intercropping systems can allow increased tree crop production without significant reduction in food production, which helps to protect household food
security.

Agroforestry or other interventions?

At another level, the question must be posed: to achieve significant household
welfare impact, why invest in agroforestry, rather than direct assistance in
water, health or food supplementation? The answer depends on the nature
of local household priorities, incentives and constraints, and the relative
costs and benefits of non-agroforestry alternatives. Some types of welfare
concerns cannot be addressed through trees (e.g., many health problems,
education), so if local communities identify these as their primary concerns,
then development resources might be more appropriately invested in these
areas.

But many household welfare goals can be addressed wholly or partially
through farm or community tree resources. The presence of a variety of
food-producing trees on the farm can provide a more effective form of food
security, at lower public administrative cost, than some alternative policies
(Cannell 1989). Minimal rural housing standards can probably be attained by
the existence of ample tree resources for construction materials, accessible to
the poor. Fuelwood availability can contribute to improved nutrition, heating
and processing opportunities.

Agroforestry and social forestry also offer the opportunity to address sim-
ultaneously other social objectives, both environmental and economic. Kenya
data suggest that trees grown for cash, production inputs or erosion control
can at the same time provide essential household goods and services, either
through multi-purpose species or mixing of different species in proportions
guided by the mix of household needs. Agroforestry can contribute to capital
formation, and programmes which provide farmers with new productive
skills in agroforestry can expand future household options without further
intervention.

Objective assessment of the alternatives requires fairly detailed information on farmers' existing uses of planted and naturally-growing trees, their
economic value, market features, non-economic constraints, and the costs
and benefits of agroforestry and non-agroforestry interventions in addressing
key welfare objectives at a community or regional level. At this time, few
agroforestry programmes have such information available.

Conclusions

The Kenya case study provides some indications of the evolving role of agroforestry in meeting household needs under conditions of land use intensification and market development. More rigorous confirmation of the relationships identified, however, was precluded by the weakness of available data about agroforestry in the region. Unfortunately, for the state of agroforestry research and development, the Siaya and South Nyanza databases are better than those available in most other places.

It is critical that agroforestry development projects, government land-use monitoring programmes, and national and international research organizations begin to collect the data which are needed to understand these processes of agroforestry adoption and diffusion. On the one hand, we need to develop time-series data on different types of tree cover, types of farm trees and agroforestry systems, and consumption, production and marketing (including seasonal price data) of key tree products. This will allow monitoring of trends over time, and provide a context for the implementation of case studies which provide more detailed analysis of adoption behaviour at the household and community levels.

The latter require linking data on tree management by households directly with agricultural land management, and with socio-economic data describing basic characteristics of the household. Data on household cash, labour and tree product use budgets should improve our understanding of the role and impact of agroforestry in the household economy, and the sensitivity of adoption behaviour to changes in these characteristics.

Depending upon the user, such studies could be implemented in a rigorous quantitative manner through household and regional surveys, or in a rigorous qualitative manner through farmer and group interviews which explicitly link tree and crop production decisions with household resources and livelihood strategies.

With accumulating experience and research, our understanding of agroforestry's contribution to rural welfare, under different socio-economic and agroecological conditions, should improve. Greater cross-programme and cross-region comparisons of the effects of technical and policy interventions, together with regular feedback from field agroforesters, should generate more reliable guidelines for national and regional policy planners. But at the local level, participatory planning processes whereby local people and communities can identify their own priorities for tree growing will probably continue to generate more reliable project agendas than top-down planning (Scherr 1992b).

Annex. Data sources

Between 1988 and 1990, the International Council for Research in Agroforestry (ICRAF) collaborated with CARE to assess the adoption and impact of agroforestry practices among farmer participants in the AEP. Studies included:

- focus group interviews with extensionists and group members;
- individual farmer interviews;
- existing AEP extension records, including group diagnosis of problems and agroforestry priorities, a baseline survey from 1985 (which provided little quantitative data on agroforestry practices), and nursery and survival records;
- a sample survey in 1988 of 126 Siaya farmers on alley-cropping and tree border management;
- a sample survey in 1989 of all agroforestry practices on 336 Siaya and South Nyanza farms;
- historical research on district land use, population and forestry from Colonial and post-Independence archives; and
- aerial photographs of the region taken in the early 1980s estimating area under different types of tree cover (Ecosystems Ltd 1984).

Principal quantitative data reported in this paper are derived from the 1989 survey. The sample for the survey was randomly selected from among all CARE-assisted group members, and stratified according to region: the higher altitude, high rainfall area dominated by very small crop farms; the medium rainfall area, also with small farms; and the low rainfall, lower altitude zone with less reliable second planting season, larger farm sizes, and greater dependence on livestock.

Assessment of farmer change in agroforestry practices was based on documented differences in numbers, uses and species of trees grown before and after CARE intervention. Theoretically, a comparison between participants and non-participants would have been preferable, but lack of an identifiable baseline for non-participants, the strong local impact of CARE on agroforestry extension by other agencies, and logistical difficulties argued against this option. The principal research interest was in identifying patterns of change. AEP intervention did appear to increase the intensity of agroforestry (for further discussion, see Scherr 1992a). Because farmers made the principal decisions about species, sites and uses, and because of broad access to germplasm from non-project sources, the resulting patterns are likely to reflect actual demand. Thus, in this paper, pre-CARE agroforestry practices are seen as a proxy for 'typical practices', while practices established during the project period indicate likely directions for intensification.

Data collected included a complete census of all trees on farms, planted and naturally-growing, with their uses, time of planting (pre- or post-project), site, configuration, associated crops, and whether benefits had yet been received. Area of all plots with trees was estimated using paced distances. Qualitative information was collected on species preferences, information needs, ranking of sources of fuelwood, poles and fruit, sale of tree products, and extension contacts, sources of planting materials, and constraints to tree growing.

Farm information included area in cropland, livestock number and feeding practices, and ecozone. Only minimal household data were collected, on wealth category and sex and status of the group member. No price, income, production or cost data were collected, and there is no historical data on agroforestry production and consumption. Thus, while the study permits quantification of changing agroforestry practices, it does not permit quantitative testing of the factors explaining change. Together with local interviews, archival information, and aerial photography, however, the data suggest principal patterns of change in agroforestry practices over time and space, and their changing role in the household economy.

Notes

1. The author would like to acknowledge the farmers and staff of CARE International in Kenya Agroforestry Extension Project who were jointly involved in much of the research reported here. Field studies were carried out while the author was a principal scientist at ICRAF in Nairobi, Kenya. Mike Arnold, Peter Dewees, Peter Hazell, Steve Vosti, and Kadi Warner provided insightful comments on earlier drafts. At the International Food Policy Research Institute (IFPRI) in Washington, DC, Patricia Bonnard provided valuable assistance in data analysis. Support to ICRAF for field research was generously provided by the Australian International Development Assistance Bureau, the Ford Foundation and the Rockefeller Foundation. Japanese International Cooperation Assistance generously supported the author's work at IFPRI.
2. These conclusions are drawn from a selection of material found in the Kenya National Archives covering the period from 1910 to 1958, particularly, Nyanza Province and Central and South Nyanza (Kavirondo) District records of Provincial Commissioners, Agricultural Officers and Forestry Officers.
3. CARE International is an international private voluntary agency that has been involved in relief work and development assistance since the 1940s. The main areas that CARE has worked in in Kenya have been community-based health, school-building, water systems, women's income generation, and social forestry. Work on the Agroforestry Extension Project, which has since become a model for more than 30 other agroforestry projects established by CARE elsewhere in the world, began in 1983. Siaya and South Nyanza Districts were selected for this pilot project for several reasons: serious malnutrition, fuelwood shortages and soil degradation; lack of cash crops and income for purchase of poles, fuelwood, fruit, etc.; a high proportion of female-headed households facing labour constraints; environmental concerns about the loss of natural woody vegetation; and Forest Department efforts to encourage farm forestry in the area.
4. The term 'free-standing tree' is used in this paper to describe trees which grow as clearly recognizable individual trees. This contrasts with the 'dense hedge' in which many trees or shrubs are grown so closely together that individual trees cannot be distinguished. In the survey reported on here, free-standing trees were counted as

a measure of quantity, while, for dense hedges, quantity was measured in terms of metres of hedge.

References

Barlett, P.F. (ed.) (1980). *Agricultural decision making: anthropological contributions to rural development*. Academic Press, New York.

Baum, K.H. and Schertz, L.P. (1983). *Modelling farm decisions for policy analysis*. Westview Press, Boulder.

Binswanger, H. and Ruttan, V. (1978). *Induced innovation: technology, institutions and development*. Johns Hopkins University Press, Baltimore.

Bonnard, P. and Scherr, S. (1994). Within gender differences in tree management: is gender distinction a reliable concept? *Agroforestry Systems*, **25**, 71–93.

Boserup, E. (1965). *The conditions of agricultural growth*. Aldine Publishing, Chicago.

Bradley, P. (1991). *Woodfuel, women and woodlots*, Vol. 1 and 2. McMillan Press, London.

Buck, L. (1990). Planning agroforestry extension projects, the CARE International approach in Kenya. In *Planning for agroforestry* (ed. W.W. Budd, I. Ducchart, L.H. Hardesty and F. Steiner), pp. 101–31. Elsevier, Amsterdam.

Cannell, M.G.R. (1989). Food crop potential of tropical trees. *Experimental Agriculture*, **25**, 313–26.

Chambers, R. and Leach, M. (1989). Trees as savings and security for the rural poor. *World Development*, **17**, (3), 329–42.

Ecosystems Ltd (1984). Western Kenya integrated land use database: maps. Ecosystems Ltd, Nairobi.

Falconer, J. (1990). Agroforestry and household food security. In *Agroforestry for sustainable production: economic implications* (ed. R.T. Prinsley), pp. 215–39. Commonwealth Science Council, London.

Gregersen, H., Draper, S. and Elz, D. (ed.) (1989). *People and trees: the role of social forestry in sustainable development*. Economic Development Institute of the World Bank, Washington, DC.

Holden, D., Hazell, P., and Pritchard, A. (ed.) (1991). *Risk in agriculture*, Proceedings of the Tenth Agriculture Sector Symposium. World Bank, Washington, DC.

Muturi, W.M. (1991). Availability and use of wood products: a case study of the high potential zone of Siaya District, western Kenya. In *Socioeconomic research for agroforestry systems development*, Proceedings of an International Workshop held in Nairobi, 2–6 September 1991 (ed. S.Franzel and M.Avila), pp. 57–73. International Council for Research in Agroforestry, Nairobi.

Oduol, P.A. and Akunda, E. (1987). Trip report to CARE/Kenya Agroforestry Extension Project. Unpublished ICRAF-CARE Report No. 4. International Council for Research in Agroforestry, Nairobi.

Rocheleau, D.E. (1987). The user perspective and the agroforestry research and action agenda. In *Agroforestry: realities, possibilities and potentials* (ed. H.Gholz), pp. 59–87. Martinus Nijhoff, Dordrecht.

Ruthenberg, H. (1980). *Farming systems in the tropics*. Clarendon Press, Oxford.

Scherr, S.J. (1985). *The oil syndrome and agricultural development: lessons from Tabasco, Mexico*. Praeger, New York.

Scherr, S.J. (1993). The evaluation of agroforestry practices over time in the crop–livestock system of western Kenya. In *Social science research in agricultural technology: spatial and temporal dimensions* (ed. K. Dvořak). CAB International, Wallingford.

Scherr, S.J. (1992a). The role of extension in agroforestry development: evidence from western Kenya. *Agroforestry Systems*, **18**, 47–68.

Scherr, S.J. (1992b). Not out of the woods yet: challenges for economics research in agroforestry. *American Journal of Agricultural Economics*, **74**, 802–8.

Scherr, S.J. and Alitsi, E. (1991). The impact of the CARE Agroforestry Extension Project. Unpublished report to extension staff. International Council for Research in Agroforestry and CARE International in Kenya, Nairobi.

Singh, I., Squire, L. and Strauss, J. (ed.) (1986). *Agricultural household models: extensions, applications and policy*. Johns Hopkins University Press, Baltimore.

Steppler, H.A. and Nair, P.K.R. (ed.) (1987). *Agroforestry: a decade of development*. International Council for Research in Agroforestry, Nairobi.

Vonk, R. (1986). Report on a methodology and technology-generating exercise: the Agroforestry Extension Project, 1982–86. Unpublished report. CARE International in Kenya, Nairobi.

Vosti, S.A., Reardon, T. and von Urff, W. (ed.) (1991). *Agricultural sustainability, growth, and poverty alleviation: issues and policies*, Proceedings of a conference held in Feldafing, Germany, 23–27 September 1991. Deutsche Stiftung für internationale Entwicklung (DSE) Zentralstelle für Ernährung und Landwirtschaft, Feldafing.

7 Farmer responses to tree scarcity: the case of woodfuel

Peter A. Dewees[1]

7.1 Tenets of orthodoxy

Introduction

Even when households have heavy demands for tree-based products, these demands in themselves may have little impact in acting as incentives for farmers to plant trees. Such an outcome is especially problematic for planners who try to develop investment programmes in order to meet projected demands for particular tree products. These types of projections tend to value the required inputs and outputs in ways which are at great variance with the ways in which rural households value them. This has been particularly so in the case of woodfuel project interventions.

A growing body of experience with woodfuel initiatives has suggested that some of the more conventional views of woodfuel supply and demand, in this chapter referred to as 'woodfuel orthodoxy', have been far too simplistic in suggesting how households in developing economies are able to adapt to constraints related to domestic energy consumption. A central tenet of 'woodfuel orthodoxy' is that woodfuel is principally scarce because there are not enough trees, and so these scarcities can be most effectively alleviated by tree planting.

We take the view here, however, that rural people seldom view woodfuel scarcities in this way, in isolation from the household's other constraints. Because of this, the seemingly clear link between supply and demand which is implicit in conventional notions of woodfuel scarcity and abundance provides little guidance for planning or project development. Indeed, this approach has often been highly costly and unproductive.

We would argue instead that, from the farmer's perspective, tree planting is not usually the most rational way to respond to woodfuel scarcities. We question the mechanisms which have been taken to alleviate perceived scarcities, and, finally, suggest that the singular focus on woodfuel production in many aid projects has obscured more fundamental issues relating to household resource allocation and factor endowments.

The literature is quite full of references to the impacts of woodfuel scarcity on households: increased labour time for fuelwood collection, a deterioration in the quality of woodfuel used, an increased reliance on agricultural residues, the emergence of markets, and so on. We suggest that there is scope for an alternative interpretation of the evidence. We argue that these impacts are far more often the outcome of much more fundamental issues related to labour use, land endowments, the transition from subsistence to market economies, and cultural practices. If indeed these more fundamental issues are involved, then the conventional response of development planners to perceived domestic energy shortages, which generally consists of support for tree planting interventions, may have little or no impact.

From the outset, it should be emphasized that it is *not* an objective to imply that there are not serious woodfuel scarcities in some areas, or that woodfuels are somehow no longer a basic human need for the great majority of rural people who rely on them as a source of household energy. Rather, it is to suggest that the problem is fundamentally quite different from the way it has generally been perceived. Consequently, the range of responses to emergent scarcities on the part of development assistance agencies has been, in large measure, inappropriate.

The 'woodfuel gap'

Anyone who tries to understand the nature of woodfuel scarcities by playing with the numbers comes very quickly up against the inevitable: the population driven demand for woodfuel cannot possibly be met by the sustainable management of existing tree resources. The conclusion is unavoidable. If these demands are to be met, orthodoxy suggests that vast areas of forested lands must be cleared in the immediate future, or a massive programme of reforestation must be undertaken. As guidance for energy policy and planning, these types of conclusions are unhelpful. Yet they continue to be used to justify often expensive programmes of tree planting and reforestation.

The Food and Agriculture Organization of the United Nations (FAO) (1981) estimated that more than 100 million people already face acute woodfuel scarcities, and that nearly 1.3 billion people live in woodfuel deficit areas.[2] While it is generally agreed that the data on which the FAO study was based were incomplete, and that it was unclear and poorly understood how people are able to respond to woodfuel scarcities, the analysis suggested that the underlying trends are inescapable: for huge numbers of people, woodfuels will become increasingly scarce and expensive.[3]

Estimates of the rates of reforestation which would be needed to meet future woodfuel demands are staggering. A widely cited World Bank study suggested, for instance, that the rate of tree planting in Africa would have to be increased 15–fold if year 2000 demands for woodfuel were to be met (Anderson and Fishwick 1984). In Malaŵi, a review carried out some years ago concluded that around 800 000 ha of fast growing trees would have to be planted to meet estimated 1990 'deficits' of 8 million m³, at a phenomenal cost of well over US$360 million (French 1986).

In the absence of any intervention, it is generally accepted that the 'woodfuel gap' between supply and demand will widen considerably. As used and expressed by policy makers, the gap has provided much of the rationale for many of the woodfuel project interventions of the last 15 years.

In some cases, the woodfuel gap has formed the basis for national energy policy and planning activities. In Kenya, for instance, household energy surveys in the late 1970s were used as the basis for a complex computerized energy information model, the Less developed country (LDC) Energy Alternatives Planning system (LEAP). This model was developed to provide a means for evaluating the impact of different energy policy and planning initiatives (Raskin 1986). The Kenya 'base case' analysis of woodfuel supply and demand, reflecting the outcome 'if prevailing conditions and practices remain unaltered for the remainder of the century', are summarized in Fig. 7.1.

Projections of energy supply and demand allow policy makers and planners to create alternative woodfuel futures by assessing the outcome of different project and energy policy initiatives. Planners can hypothetically attempt to close the gap through these initiatives. By justifying specific proposals on the basis of their impact on closing the gap, aid agencies can be drawn into the scheme of things by providing finance for these initiatives. In Kenya, the hypothetical 'policy case' which closed the woodfuel gap through different

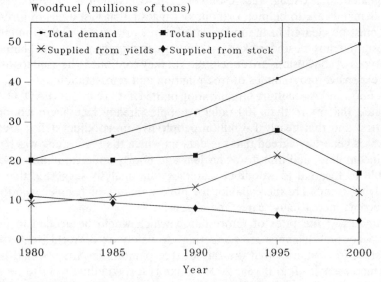

Fig. 7.1 Kenya 'base case', projected woodfuel supply and demand gap, 1980–2000 (from O'Keefe *et al.* 1984).

project interventions, described woodfuel planting and tree management activities targeted for around 3.6 million ha, as well as a range of demand management strategies (O'Keefe *et al.* 1984).

There are admittedly serious woodfuel shortages in some areas of Kenya, and responses to scarcity may involve very high economic costs. Analyses of the woodfuel gap, however, tend to obscure the scope of the problem. Indeed, the LEAP analysis leaves more questions than it answered. What is the basis, for instance, for concluding that woodfuel supply equalled demand in 1980? What indicators of scarcity did LEAP researchers use? Could there in fact have been a woodfuel gap in 1980? Conversely, is there the possibility that woodfuel use in 1980 was *far greater* than that which would have been necessary to meet basic human needs?

Woodfuel supply/demand analyses can be quite misleading because they fail to distinguish between demand, latent demand, consumption, and the extent to which current demands (or year 1980 demands in the Kenya analysis) can be acceptably moderated without seriously affecting the household's access to basic human needs: adequate food, warmth, housing, clean water, education, and so on. LEAP projected a woodfuel gap based almost exclusively on current consumptive trends. Rather than suggesting that people in Kenya might moderate their demands for woodfuel as a response to scarcity, the LEAP 'base case' analysis projected that demands for woodfuel would actually *increase* from 1.3 tons per capita in 1980 to 1.5 tons per capita in the year 2000. Compared with current rates of consumption in other countries, this must surely be among the highest rates of per capita consumption.

This Malthusian woodfuel gap perhaps best represents the chasm in our thinking about the dynamics of woodfuel supply and demand, and about the ways rural people go about responding to woodfuel scarcities. The bias inherent in most analyses is that aid-financed project interventions are the solution. In many areas, however, rural people are *already* responding to increased woodfuel demands in ways which are innovative and imaginative and which involve far lower economic costs than many project interventions.

Measures of woodfuel scarcity and abundance

One of the things which should be manifestly evident from the 'conventional' Energy Crisis of the 1970s was that there was tremendous scope for reducing the consumption of fossil fuels. In 1984, for instance, energy consumption among the industrialized countries was *still* considerably less than it was before the 1973 oil price shocks.[4] Few people would argue that this reduction in consumption has been achieved at any great long-term social or economic cost. Many would argue that the level of energy consumption remains unacceptably high.

In the mid-1970s, however, the 'present trends' sort of analysis of conventional energy supply and demand (a type of analysis which has continued to characterize the woodfuel debate) obscured the likely demand response to changing prices. As Singer (1984) in his critique of the Council on Environmental Quality's (1982) estimates of conventional energy demand suggested:

The projection of future energy demand has become a popular pastime, consuming much effort and resources. It is, of course, motivated by people's desire to know whether there will be a 'shortage', i.e. whether 'demand will exceed supply'. To those who believe that prices in a free market can allocate available fuels efficiently, and bring forth the necessary resources, the idea of a long-term shortage makes no sense whatsoever; the projection of energy demand without considering price changes should be viewed as an academic exercise without much policy content . . . . [Energy projections] have been misused by politicians to create the spectre of a 'crisis', and have served to involve the government in dubious interventions into the energy market.

Similarly, it is a fallacy to suggest, particularly in global analyses, that present trends in woodfuel consumption are likely to continue, or that policy and planning initiatives should be geared to maintaining these sometimes quite high levels of consumption. In the broadest sense, woodfuel consumption is closely linked to its economic cost. Possibly radical changes in patterns of consumption can be anticipated as its cost increases. Indeed, woodfuel consumption is quite dynamically related to its economic cost and to supply. It is not a static variable which can be extrapolated from population growth rates.

Consider, for instance, data from woodfuel surveys carried out in the late 1970s. Woodfuel consumption in Papua New Guinea was reported to be around 1.8 m^3 of woodfuel equivalents per capita. In Nepal, the figure was estimated to be around 0.9 m^3 per capita, while in Afghanistan, it was estimated to be around 0.3 m^3 per capita (Agarwal 1986).

These data suggest the not-surprising conclusion that woodfuel consumption is a function of the cost of obtaining it. In arid Afghanistan, woodfuel consumption is likely constrained because it is difficult to obtain. It *can* be obtained, but at some economic cost: perhaps by spending additional time in collecting it, perhaps through the marketplace, perhaps from farm trees or the use of agricultural residues, or by deferring consumption to the future through changed cooking and heating habits. Data from Papua New Guinea and from Nepal suggests that woodfuel is more available, and that its economic costs to the household are much lower than in Afghanistan. If woodfuels were seriously scarce, people would not be burning as much as they are.

Even this approach is misleading, as it obscures the impact of other variables on the essentially dynamic relationship between supply and demand, and on the household's need for energy. Fisher (1979) pointed out that an adequate measure of resource scarcity should have the property of being able to summarize 'the sacrifices, direct and indirect, made to obtain a unit of resources'. While this notion suggests several admittedly more accurate measures of resource scarcity, the problem of measurement for woodfuels becomes difficult for a variety of reasons. These are associated with their often common property nature, the cost of alternatives (both in terms of capital, labour, and land use), and possible improvements in woodfuel harvesting, extraction, and delivery.

Energy constraints to development must be evaluated by balancing energy consumption with other variables and costs which influence demand. There may be, for instance, income effects: per capita woodfuel consumption may increase (or decrease) as the household's endowments of physical and human capital increase. There are likely price effects: as the price of woodfuel increases, demand may decrease (or it may increase if, as economists say, it is a 'Giffen' good). Finally, there may be substitution effects: as the price of woodfuels increase, cheaper substitutes may be found; conversely, as the prices of alternative fuels increase, demands for woodfuels may increase.

7.2 Evaluating the impacts of woodfuel scarcity

The scope for responding to scarcity

As a result of woodfuel scarcities, it is generally argued, the consumption of a marginal unit of woodfuel will entail sometimes quite considerable costs. The outcome of woodfuel scarcities includes:

- increased time for fuelwood gathering, often with implications for labour availability among women who are principally responsible for its collection;
- a deterioration in the quality, and in the type, of domestic energy used, and an increase in the use of agricultural residues and animal dung for household cooking and heating;
- an increase in deforestation because more trees have to be felled to meet greater woodfuel demands;
- changes in cooking and heating habits;
- the emergence of woodfuel markets, and where there are already markets, increases in the prices of woodfuels.

We question whether these impacts are clearly the outcome of woodfuel scarcities, or whether they are the outcome of much more fundamental issues related to the ways in which rural economies in transition operate, with their underlying culturally derived influences on land, labour, and capital allocation. If indeed more fundamental processes are involved, then the conventional response of development planners to perceived domestic energy shortages, support for tree planting interventions, may have little or no impact.

Where there *are* clear woodfuel scarcities, manifested by these types of impacts, tree planting as a mechanism for responding to them may entail particularly high costs for farmers. Without negating the clear impact on the poor of changes in rural factor endowments, brought about by woodfuel scarcities, rural people sometimes view the impacts of scarcities from a fundamentally different perspective than the way development planners perceive them. The objective here is to suggest that the maintenance of adequate levels of domestic energy consumption can be (and indeed often is) achieved at particularly low cost.

Indeed, farmers are often able to adapt to scarcities of different resources in 'positive-functioned' sorts of ways. As Dirks (1980) pointed out in his study of human responses to food shortages, adaptation to deprivation can involve both 'progressive' and 'recursive' changes. In cases where there may be risks of food shortages, these progressive responses include food sharing, shared cooking, increased labour use for food production, rationing, the use of famine foods, and migration.

People who depend on woodfuel have developed similar types of responses when it has become scarce. These responses include fuel sharing, shared cooking, increased labour-use for fuel collection, the use of other fuels when desired types of woodfuel are scarce, and migration and nomadism.

Where woodfuel scarcities are in evidence, much debate has been initiated about the impact of urban and industrial woodfuel markets on increased scarcity in rural areas. Urban woodfuel markets, though, are a special case and few generalizations about their impact on woodfuel scarcity can be made because painfully little is known about how they work. Planners often forget, as well, that woodfuel markets often provide a critical source of employment and income. In Kenya, for instance, studies have suggested that (in 1985) urban charcoal markets provided full-time jobs for around 30 000 people as charcoal burners, transporters, and retailers (Dewees 1985).

It is also sometimes argued that cash crop farm forestry contributes to rural woodfuel shortages and to increased disparities in labour use and poverty (Shiva *et al.* 1982). It is, however, as likely that farmers undertake to supply markets for woodfuel and for other tree products as a response to much more serious constraints, such as the need for rural housing, capital constraints, the failure of rural labour markets, and distortions in rural land markets.

Evaluating the extent to which woodfuel may have become more scarce requires a dynamic view. Saying that women spend, say, 3 h/day collecting woodfuel is entirely different from saying that women are spending 3 h/day collecting woodfuel, but that 10 years ago, they only spent 1 h/day (or possibly 5 h/day!). The essential improvement or worsening of the woodfuel situation must be measured within some objective time horizon.

Woodfuel scarcities and labour

A household's levels of woodfuel consumption are greatly dependent on the amount of labour that is available to collect it. As competing constraints are placed on the household's available supply of labour, woodfuel consumption can decline. These constraints become greater as woodfuel becomes more scarce because farther distances have to be travelled to collect it. They perhaps become most serious when, as Cecelski (1987) points out, fuel gathering can no longer be combined with other work (such as the collection of medicinal herbs, or as an activity on returning from the fields), but instead has to be the object of a separate trip.

The question here, however, is not clearly one of fuel availability, but one of *labour* availability. Even if fuel is available in abundance, if there are constraints to the household's supply of labour, the cost of burning woodfuels can be quite high and, consequently, consumption can be quite low. Conversely, if labour is abundant, the time spent on fuel collection and the level of woodfuel consumption can be quite high. In Zimbabwe, for instance, it has been found that the frequency of fuelwood collection, as well as the time spent for collection, increases during the dry season primarily because households are freed from agricultural labouring (du Toit *et al.* 1984). During the planting season, households spend around 3 h/week in fuelwood collection; during times of lowest labour demand, they spend around 10 h/week.

In the Sudan, there are indications that competing demands for labour may have some impact on the price of charcoal delivered to urban markets. Labour costs typically account for around a fifth of the delivered price of charcoal. Agricultural labour and labour for charcoal burning are nearly perfect substitutes. Because of the far greater demand for agricultural labour, however, the price of farm labour dictates the wage of charcoal burners (el Faki Ali 1985). An analysis of charcoal prices in Khartoum over the last 10 years suggests that charcoal prices are indeed highest when demands for agricultural labour are highest, for instance, during years when there have been particularly good harvests. In constant prices, charcoal prices in Khartoum showed peaks during the 1977 and 1981 harvests, but bottomed out during the 1984 drought (Dewees 1987).[5]

The lack of adequate on-farm labour can be an important incentive for farmers to plant trees. In Machakos District of Kenya, smallholdings headed by very young or very old women face serious labour constraints. Households headed by mothers of very small children and older infirm women have expressed particular interest in concentrating fodder and fuel resources on or around the farm so they are more accessible to the household (Rocheleau 1985). These kinds of constraints in other areas of Kenya have led to initiatives which have increased the household's access to woodfuel. In some areas of Kisii District of Kenya where there are agricultural labour shortages, woodfuel collection requires a relatively small proportion of the household's time because woodfuel supplies are obtained from planted and managed on-farm trees (S. Orvis, personal communication).

Most analysts tend to assume that fuelwood collection is principally a woman's task. The evidence suggests that the gender division of labour is not always so straightforward. Studies of household labour-use differentiated by task, sex, and age in Java and Nepal, for instance, provide startlingly different information about the role of men, women, and children in fuelwood collection. In villages surveyed in Java, children under the age of 15 provided nearly 70 per cent of all the labour required for fuelwood collection, while in Nepal, they provided only 30 per cent of the total labour required. In Java, men and boys provided 74 per cent of the labour required for fuelwood collection, while in Nepal, they provided 57 per cent. Finally, in Java, fuelwood collection accounted for 3.5 per cent of all work done, while in Nepal, it accounted for only 1.5 per cent. In absolute terms, around twice as much labour was used for woodfuel collection in Java, across all age groups, as in Nepal (Nag *et al.* 1978).

Clearly, the issues surrounding labour use and woodfuel scarcity are far more complex than most analyses would suggest. Labour constraints to woodfuel collection are labour constraints to the farm. Where farm labour is scarce, woodfuel collection may be costly; where farm labour is abundant, woodfuel collection may involve very low costs.

Without comparing labour use for woodfuel collection with labour use on the farm as a whole, it is inaccurate to imply that increased use of time for woodfuel collection necessarily reflects woodfuel scarcities. If labour is abundant, how much of a problem is it, really, if the household is spending three times as much labour in woodfuel collection than it otherwise would (as in Zimbabwe)? If labour is scarce, the impacts of scarcity are felt in all farming activities. The question perhaps becomes one of whether or not the household's access to woodfuel can be increased by reducing its labour requirements for other farming activities.

Woodfuel scarcity and the quality of alternative fuels

It is usually argued that, as woodfuels become more scarce, households are forced to rely on low quality fuels: sticks, twigs, agricultural residues, and animal dung. While a shift to lower quality fuels can indeed reflect the fact that particular types of woodfuel are becoming more scarce, this is not always the case. These fuels may be in more common use when they are preferred fuels, or when they are available at low cost and in relative abundance.

In circumstances when woodfuels are genuinely scarce, and when tree planting is encouraged as a response to fuel scarcities, trees often assume much higher economic values to the farmer; consequently, tree planting may have little or no impact on reducing the household's reliance on agricultural residues. Trees which produce 'high quality' fuelwood may be more highly valued in the rural economy, for instance, as sources of building timber, or as sources of fruit and fodder. Project interventions which have the objective of producing high quality woodfuels may be frustrated in achieving this objective because of the economically low value of woodfuels *vis-à-vis* its alternative uses.

Markets can provide powerful incentives to plant trees (Chapter 8, this volume). Prices greatly influence whether planted trees are used for woodfuel or for other goods. In Gujarat in western India, for instance, the Forestry Department's efforts at increasing supplies of woodfuel basically resulted in the establishment of huge areas of eucalypt (*Eucalyptus* spp.) woodlots for building poles. The objectives in the original project documents were clearly to produce fuelwood. Perhaps it would have been more appropriate to encourage farmers to grow scrubby shrubs which could not be used for pole wood. In fact, woodfuel *is* being produced in Gujarat, but only as a wastewood by-product of building pole production. As much as 30 per cent of the increment is non-marketable as construction poles, and is ultimately used under the pot (Dewees 1983). Overproduction for the pole market, as we discuss in Chapter 8, has brought about the felling of eucalypt woodlots for sale or use as woodfuel. The principal incentive for establishing these woodlots, however, was arguably the prospect of cash income from the sale of building poles, and not for woodfuel.

Although twigs and sticks are not, in the usual view, high quality woodfuels, from the perspective of the farmer who must rely on them, because of their abundance, they can be highly valued. In Indonesia, it has been estimated that low quality fuels are around four times more available than high quality stemwood (Weatherly 1980). Indeed, in some areas, crop residues are a preferred type of fuel, both because of their availability and because of the quality of the flame. In China, for instance, straw (which accounts for over 40 per cent of domestic energy used in rural areas) produces a hot flame appropriate for frying foods quickly (Wu and Chen 1983).

Among the Khan Magar of north-west Nepal, Molnar (1981) suggested that the use of agricultural residues as a household fuel was not necessarily related to physical woodfuel shortages. She noted that stalks of *puwa*, a nettle which is stripped for its fibre, are commonly used for fuel. Despite a heavy reliance on this substandard fuel, she observed that the forest resource base was reasonably intact, and that in the short term at least, it would be sufficient for providing adequate woodfuel and building materials.

Pant (1935) however observed that in the Kumaon Himalayas of Nepal, fodder and fuel were in seriously short supply, and that households had to rely on stalks of amaranth, hemp, and chili for fuel. He noted that these scarcities were most serious in irrigated valleys and required periodic, lengthy travel to distant forests in search of fuel and fodder. It is interesting to note that his observations were made over 50 years ago at a time when the government of Nepal was actively encouraging the conversion of forested lands into farmlands (Blaikie and Brookfield 1987). If there were shortages of woodfuels in this area 50 years ago, one could anticipate that responses to scarcities have, since then, become integrated into the local social and economic fabric. Of particular concern are those circumstances where these types of longer term adjustments to woodfuel constraints have been locally problematic.

Sometimes farmers manage woody biomass *specifically* to produce low-quality woodfuel. In western Kenya, for instance, a woody shrub (*Tithonia diversifolia*) which is widely found on verges and along paths is managed on a sustained yield basis in a way which produces large quantities of woodfuel. The bushes are cut just above the ground, and the branches are left to dry for a few days before collection. Although the productivity of these bushes has never been measured, yields are clearly quite high. In some areas of India, similar types of shrubs are grown for woodfuel, for fodder, and for their soil restoring abilities. *Sesbania aculeata*, for instance, is a leguminous shrub which is grown as a fallow crop primarily for woodfuel (Pathak 1980).

In the Philippines, smallholders were encouraged through project interventions to grow *Leucaena leucocephala* to provide fuelwood for the local tobacco curing industry and were provided loans and seedlings to enable them to do so. In the first 3 years of the project, less than 40 people had signed up for the programme, and planting achievements were less than 6 per cent of targets (Hyman 1982). Closer examination revealed that farmers were *already* growing very large quantities of woodfuel in small, very intensively managed woodlots of *Gliricidia sepium*. These woodlots were hardly noticed by project planners, probably because they produced a tangled, dense mass of 'low quality' woodfuel. When farmers were given the opportunity and were encouraged to grow high quality woodfuel, they chose to stay with their traditional and effective tree management practices (Wiersum and Veer 1983).

The burning of agricultural residues is generally thought to contribute to declining soil productivity. This is not clearly the case. Some agricultural residues which make perfectly good woodfuels cannot be dug back into the soil in a way which improves the soil structure or fertility. Stalks from cotton, cassava, pigeon peas, and chick peas, for instance, are much too woody to decompose sufficiently in a way which would improve soil productivity, and in many areas, they are used as a valuable source of fuel (Barnard and Kristoferson 1985). Indeed, there are cases where, by digging in residues, the farmer can encounter serious constraints to crop production. As residues decompose, crops and micro-organisms compete for the same nutrients (particularly phosphorus and manganese) which sometimes results in lower yields than if residues had not been dug in (Loutit and Brooks 1970; Benians and Barber 1974). Residues with a high carbon/nitrogen ratio can bring about the depletion of soil nitrogen as a result of decomposition. Consequent shortages of nitrogen, while causing lower crop productivity, can also aggravate the severity of various plant diseases (Lynch and Poole 1970).[6]

In some tropical areas, particularly where farmers are unable to practise crop rotation, plant diseases can be spread by digging in crop residues. Fungal decomposers of crop residues, for instance, can produce substances which are phytotoxic to crops. Garret (1970) concluded that the most successful way of limiting the spread of pathogens is by the destruction of infected crop residues after harvest.

The digging in of crop residues and manure is also labour-intensive. Households with labour constraints may find this to be an inefficient use of scarce resources.

With regard to the use of animal dung as fuel, there is only limited evidence which suggests that if woodfuels were more available, the use of animal dung for fuel would decline. Animal dung is most widely used as fuel in India. In many areas of Latin America, Asia, and Africa its use is virtually unheard of. Even in Nepal (where the opportunity cost of using dung as fuel was used as the basis for the economic justification of the World Bank's forestry project investments there), the use of dung as fuel is the clear exception, even in the most woodfuel scarce areas, rather than the rule (J.G. Campbell, personal communication).

Dung is clearly used for cooking in other areas (Ethiopia, Pakistan, Lesotho, Yemen, and Turkey, for instance), but the underlying issue is whether or not its use is increasing as a result of scarcities of other fuels. If instead it has been a long-term feature of household fuel use, preferred over other fuels, it may well have been integrated into the cultural tradition. The question is fundamental, because it suggests that tree planting in these situations would have little impact on reducing the use of dung as fuel.

Particularly in India, many people feel that dung has qualities as a fuel which are not shared by woodfuels. It is slow burning and produces a hot flame. It is easily transported, and is easily collected in areas where zero-grazing has become the norm for livestock management. It is often more accessible to the landless than woodfuel. Some observers have stressed the special appropriateness of dung for the preparation of ghee, and for the efficient deployment of female household labour, and it has broad significance in Hindu tradition (Lewis and Barnouw 1958; Simoons 1974).[7]

In light of the strong religious and cultural significance of the use of cow dung in India, it can indeed be argued that even if woodfuels were more widely available as a result of tree planting interventions, households would continue to rely on the use of dung as a domestic fuel.

Woodfuel scarcities and deforestation

Links between woodfuel use and deforestation are usually discussed from two perspectives: first, woodfuel consumption is often identified as an underlying cause of deforestation; and secondly, in areas which have been deforested, woodfuels are thought to have become increasingly scarce. Neither of these observations clearly describe the norm, although they have been widely accepted and form the rationale for many tree planting project interventions.

Despite a continuing emphasis on the contribution of woodfuel consumption to deforestation, it is becoming increasingly accepted that the primary causes of deforestation are more closely related to land clearance to support agricultural expansion (Bajracharya 1983).

Livestock pressures, as well, may limit the regenerative capacity of dryland forests. Where livestock grazing has intensified, and where traditional strategies of communal lands management have broken down, soil and rangeland degradation has sometimes been accompanied by a deterioration in the stock and quality of woodfuel. Tree planting however, begs the question. The problem is more fundamentally related to range management than it is to a shortage of trees.

Woodfuel harvesting by itself is not necessarily a destructive form of tree management. The problem with agricultural expansion and with land clearance is usually that whole trees are uprooted. Fuelwood collection, on the other hand, may involve hacking off a few branches from a live tree, or the collection of dead wood. As long as the root stock is not destroyed, the productivity of woodlands under this type of stress, brought about as a result of woodfuel harvesting, can be significantly higher than woodlands which are not stressed this way.[8] Woodfuel collection, rather than having a serious and destructive impact on tree cover, can often be a productive tree management strategy.

Sometimes tree management is passive. Farmers may make few conscious choices about the ways trees are used for woodfuel production. As pressures on the remaining resources become more intense, management strategies may become much more active. Active management strategies include rotational harvesting, the protection of naturally-regenerating seedlings, the protection of specific types of valued trees during land clearance, and so on. The most active of tree management strategies involve tree planting and sustained-yield management (Klee 1980; FAO 1986).

Although woodfuel demands may sometimes contribute to deforestation, woodfuel scarcity (and the need for responding to scarcity) can become most critical as an outcome of deforestation. As Eckholm *et al.* (1984) note, '[Fuelwood] scarcity is as much a consequence as a cause of deforestation. First, the widespread clearing of lands for agriculture severely reduces the available forest area. At that point, the gathering of fuel from the remaining woodland may well begin to exceed the sustainable harvest.'

Can it be concluded that farmers in areas which are heavily deforested are likely suffering from serious woodfuel shortages? Not necessarily. Indeed, the extent of deforestation may be a particularly poor indicator of the severity of woodfuel scarcities, simply because farmers may well have developed sophisticated tree management strategies, as well as cultural responses, to enable them to deal with scarcities.

In Zimbabwe, recent studies have indicated that, despite rapid rates of deforestation, residual woodlands and other tree resources are being managed to meet local needs. Even in the most heavily deforested areas, where remaining tree resources are essentially confined to non-cultivable locations (particularly hilltops), the clearance of residual woodlands over the last 15–20 years has been insignificant, despite rapidly increasing demands for these resources. In fact, some areas of residual woodland appeared to be more densely forested now than before, possibly because of coppice regrowth (du Toit *et al.* 1984; Campbell and du Toit 1988).

Over half of the respondents interviewed in field surveys in Zimbabwe reported that fuelwood was easy or fairly easy to obtain, while nearly 70 per cent believed there is currently enough fuelwood to meet household demands (an interesting distinction between *economic* and *physical* scarcity).[9] Despite the fact that the deforestation of savannah bush has clearly accelerated over the last 20 years, farmers noted that, even in the most heavily deforested areas, fuelwood is not necessarily difficult to obtain.

Similar findings with regard to the consumption of wild fruit in Zimbabwe are reported by Campbell (1987). Deforestation does not appear to affect the availability of wild fruit, because people tend to protect preferred fruit trees while land is being cleared for cultivation. Indeed, in some agricultural areas forest cover has been manipulated to such an extent that fruit trees have become the dominant type of tree. Other researchers have since confirmed that in some areas of Zimbabwe, farmers have a very sophisticated notion of the interactions between trees and crops and are clearly aware of a broad range of management possibilities (Wilson 1989; Scoones 1989; McGregor 1991).

In West Africa, tree management is believed to have been one of the earliest types of farming practices among settled agriculturalists. Evidence of the manipulation of forest cover dates back some 5 000 years. In south-eastern Nigeria, dense stands of oil palms ('oil palm bush') within tropical rain forests are more recent indications of the earlier integration of trees in farming systems. When the forest was originally cleared for farming, oil palms were left in fields because they were a valued source of cooking oil. After farm sites were abandoned, oil palms became the dominant species in the climax tropical rain forest (Sownmi 1985). The British explorer, Mungo Park, in the 1790s made careful reference to the economic importance of the shea butter tree in Mali, and noted that it was seldom planted, but that 'in clearing wood land for cultivation, every tree is cut down but the shea' (Park 1799).

Similar types of tree management strategies have been widely reported elsewhere. Even so, many development planners maintain an overwhelming prejudice about the ability of rural people to manage their environment, despite the fact that environmental management is often fundamental to their survival. These prejudices include the notion that farmers fail to understand the long-term and intergenerational impacts of their choices about resource use.

In his analysis of woodfuel scarcities and abundance in the Sahel, Foley (1987) pointed out that although energy planners at the national level often envisage a woodfuel crisis of catastrophic proportions (and respond accordingly), at the village level, farmers seem to have few worries about woodfuel shortages or to show any interest in planting the tree seedlings provided for them. Sometimes local people are blamed for their lack of foresight or their 'inability to understand exponential growth'. It is more likely that, given the community's land and labour resources, and their access to bush, to managed fallow land, and to on-farm trees, they have developed quite responsive means of dealing with growing demands for tree resources.

The impact that tree planting programmes would have on reducing woodfuel scarcities caused by deforestation is unclear. To begin with, deforestation does not necessarily bring about scarcities of woodfuel, and even when it does, farmers may have alternative and sophisticated means of responding to scarcities through environmental management strategies or through cultural practices which require no project intervention.

A final point should be made about adaptive strategies in areas which have long been deforested (or which perhaps never supported any forests). In a study of energy flows in households on the north-eastern shore of Lake Titicaca in Peru, Collins (1983) suggested that seasonal migrations are a response to energy scarcities. Seasonal migration and the development of very low-energy intensive cooking strategies also characterize the cultural and economic practices in energy scarce areas of the Himalayas.

There are interesting migration paradigms in the literature about human responses to the scarcity of other resources. The entire population of the great Moghul sandstone city of Fatepur Sikhri migrated as the result of shortages of water. Dirks (1980) noted that migration is often an outcome of food scarcity. Others have noted that the 'most obvious pastoral adjustment to a scarcity of resources is to move elsewhere; nomadism itself is created by such a necessity' (Swift 1977).

Woodfuel scarcities and changed cooking habits

When woodfuel scarcities have become most serious, woodfuel consumption can be reduced, but sometimes at a very high cost to the household. Woodfuel scarcities may result in the preparation of fewer meals or may bring about changes in the diet which favour fast-cooking, and possibly less nutritious foods. While it is widely argued that changed cooking and dietary habits are an outcome of woodfuel scarcity, studies which have supported this view are scarce and are mostly anecdotal. The issue is somewhat complicated by the fact that changes in food consumption as a result of fuel scarcity may easily be confused with changes in consumption as a result of food scarcity. The vast literature about social and cultural aspects of diet and nutrition suggests that other constraints likely play a far greater role in causing nutritional deficiencies.[10]

There are numerous low-cost strategies for reducing fuel consumption when wood is scarce, and these may involve no change in food consumption. Fire management is the most obvious low-cost means of reducing fuel consumption. Studies have shown, for instance, that the efficiency of a three-stone fire can be quite high, if the fire is closely tended and managed. The first outcome of woodfuel shortages will likely be a change in cooking habits, a change which can be accomplished at very low cost and which may have no impact on food consumption.

Changed strategies of preparing food can also reduce fuel consumption. The soaking of foods such as beans and lentils can greatly reduce cooking time. In the high altitude Khumbu Valley in Nepal, women quite typically use pressure cookers (left by numerous mountaineering expeditions) to reduce their cooking time. The adoption of metal pots and the abandonment of clay pots in some parts of West Africa was a result of convenience, as well as the fact that food can be cooked more quickly in metal pots.

In the Sudan, one outcome of growing demands for woodfuel is the 'shared pot' where women from several households may cook together, to reduce their individual woodfuel requirements. Communal cooking has become an important social focus for women in small communities (T. Hammer, personal communication). Household size can greatly influence levels of per capita woodfuel consumption. As household size increases, consumption per capita will decrease.

The concept of the communal sharing of woodfuel resources has similarities to other cultural responses when food scarcities threaten. At the prospect of having to deal with food shortages after a typhoon, villagers on the island of Tikopia in Micronesia increased the extent of food sharing, and developed the concept of the 'linked stove' (Schneider 1957).

A comprehensive review of wood conserving stove programmes (Foley *et al.* 1984) concluded that the role of improved stoves was limited.

[Because] stoves are inefficiently used and deteriorate, because wood is burnt for reasons other than cooking, and because improved stoves cannot be got to everyone, national wood savings through stove programmes can never be significant. However, though improved stoves may not save trees or forests, they can improve the daily lives of human beings.

Similar conclusions can be reached about the role of tree planting programmes with respect to human nutrition. There are likely lower-cost ways of creating additional energy supplies by conserving woodfuel. When conservation is no longer an option, other strategies can be undertaken to augment supplies. At that point, tree planting may make a great deal of sense to the household, but for entirely different reasons: because of the potential for producing construction timber, fruit, fodder, shade, and so on.

Woodfuel scarcities and the emergence of woodfuel markets

It is widely argued that the emergence of woodfuel markets is an indicator of scarcity and that where there are woodfuel markets, increasing prices are a further indication of scarcity. Both of these arguments seem to be based on misconceptions about the role of trade and exchange in the transition from subsistence to market economies.

The presence of woodfuel markets is no indicator of widespread physical scarcity and is a tenuous indicator of economic scarcity. The emergence of markets for woodfuels, and for other commodities, is an outcome of the process of specialization and exchange. As Brumfiel (1980) pointed out in a review of markets in the Aztec state,

Specialization allows for exploitation of differences in the natural abilities of individuals and in the natural resources of geographic regions. It permits economies of scale and minimizes investments in duplicating the tools of production. Exchange provides for the transfer of goods and services . . . [Exchange] is therefore essential if the benefits of specialization are to be realized.

In the Valley of Mexico, environmental diversity was associated with a degree of economic specialization. Many communities were supplementing subsistence agricultural activities with other activities which produced surpluses for the market.

Similarly, the emergence of woodfuel markets is an outcome of the need for greater degrees of economic specialization and exchange. Woodfuel markets may become a feature of the economy, for instance, when there is wage employment. This type of specialization means that some households may not be able to collect fuelwood, and may have to rely on the market to provide it for them. The fact that there are woodfuel markets gives little indication of its *economic* costs, which may be exactly the same in the absence of markets: it may still take, say, a half day to collect and chop a load of fuelwood, regardless of whether or not there are markets.

Using the 'food/fuel' paradigm, one could argue that the presence of markets means that woodfuel-selling households are generally having their needs met. In the Tudu region of Niger, for instance, no grain enters the market until households have 18 months supply in reserve (Faulkingham 1978). Similarly, a household which is selling woodfuel must be confident that it has an economic surplus: the value of surplus woodfuel to the household is greatest only if it is sold, and the household would otherwise have a means of doing without.

Woodfuel pricing trends are similarly unconvincing indicators of physical scarcity. An analysis of longer term woodfuel pricing trends in urban markets in South Asia indicated there were no convincing links between the extent of physical tree resources and market prices (Leach 1987). In the Sudan, there are large seasonal price fluctuations, but over the last 10 years, real prices have shown no long-term increases (Dewees 1987).

There may in fact be growing physical scarcities of woodfuel in these areas, but pricing trends provide us with little information about their extent. The data instead suggest the lack of an *economic* scarcity of woodfuels. In Kenya, for instance, woodfuel prices have been kept low because of improvements in the transportation infrastructure, because of an extremely competitive market, and because of the continuing clearance of agricultural land.

In Sudan, charcoal markets have become extremely sophisticated and show a high degree of vertical integration. A few entrepreneurs, for the most part, control the major aspects of production, transport, distribution, and sale, thus keeping overheads low, and allowing them to maintain their margins in the face of increasing physical scarcity.

Increases in relative prices indicate economic, and not physical scarcity. The passenger pigeon, for instance, was first commercially harvested in the 1840s, but became extinct in the 1890s. Market prices showed little tendency to increase over time, even in the face of extinction. Prices failed to rise (and hence we can argue passenger pigeons were not economically scarce) because they were a common property resource, because there were cheap substitutes, and because there were improvements in harvesting technologies (Howe 1979). It can be convincingly shown that woodfuel pricing trends similarly give no adequate indicator of physical scarcity for precisely the same reasons.

Until conventional measures of economic scarcity more accurately reflect the physical scarcity of woodfuels, the augmentation of woodfuel supplies for the market through tree planting project interventions will be problematic. In the absence of any market intervention, it is unlikely that sustainably produced woodfuel will be able to compete in the market-place with non-sustainable production, unless it is produced as a by-product of other more lucrative tree growing practices. Otherwise, tree growing solely for woodfuel production will only become profitable *vis-à-vis* alternative land and labour uses as absolute physical scarcities emerge, and as prices for conventional fuels become too high.

This is not an argument for intervention in woodfuel markets. Good market intelligence about woodfuel markets is limited to a handful of studies, and most of these were one-off efforts which fail to capture the long-term supply/demand dynamic. It would be fatuous to suggest that they could form the basis for realistic and effective interventions.

7.3 Responding to tree-based constraints to development

There is a huge gap in our understanding about the difference between the *economic* scarcity of woodfuel and its *physical* scarcity. *Physical* scarcity refers simply to whether woodfuel resources are physically present or absent. *Economic* scarcity refers to the ability of a household to allocate its land, labour and capital resources in a way which enables it to actually use this woodfuel. A resource which is physically scarce may not be economically so. Conversely, a resource which is economically scarce may not be physically so.

For example, a household may be quite remote from forests or woodlands or may have few trees growing around the homestead, yet these physical scarcities might be alleviated if capital is available to buy woodfuel in the market. Even though a household might live close to woodlands or forests, if tenure to these resources belongs to some other individual or group, or if they have other more highly valued uses such as for browse or for building material, as woodfuel resources, they are economically scarce.

Even when woodfuels have become physically scarce, households have a great deal of latitude in developing adaptive responses (which may include tree growing or management). From the farmer's perspective, the impacts of economic scarcity may not be nearly as serious as the extent of physical scarcity might suggest. It would be unrealistic, for instance, for a household to use kerosene or other fossil fuels as woodfuels become physically scarce as long as there is a broad range of much lower-cost options open to it, such as conservation. Eventually, as the economic cost of these other options increases, scarce household income may well have to be spent on kerosene.

Even in the absence of physical scarcity, there may be an economic scarcity of woodfuel, defined by the household's access to labour, income, or land. The point is simply that even if estimates of supply/demand balances suggest that households are experiencing serious woodfuel shortages, this fact alone gives no indication of whether or not these perceived 'shortages' will provide an incentive to encourage farmers to plant trees. Competing demands for households' capital, land and labour resources may limit the extent to which they are able to respond to woodfuel scarcities by increasing supplies through tree planting.

The 'woodfuel crisis' was a powerful catalyst in the late 1970s for focusing the attention of development planners on some of the more fundamental requirements of the rural poor. Too often, this myopic focus was the driving force behind project investments. Even though planners would claim that they were also interested in food, fodder, housing, and so on, an overwhelming emphasis on woodfuel emerged, probably because it seemed so much more tangible than other aspects of tree resource use. It is difficult to imagine, for instance, that much excitement could have been generated over a 'fodder crisis' or over a 'fruit crisis'.

Even much more obvious constraints such as rural housing have often been ignored during the design of forestry project interventions because of a focus on fuelwood. When projects have the objective of providing the resources to enable farmers to produce fuelwood, and they instead grow building poles, project planners are often loudly criticized because the project had not met its original objectives. The point is, from the farmer's perspective, building poles make much more sense. No one thought to ask if there were rural housing shortages.

Tree resources are of fundamental importance to farming systems, and rural people may become much more involved in tree planting or management when trees become scarce. Woodfuel scarcity in the economic sense, however, cannot always be equated with tree scarcity. So tree planting interventions which are intended to be responsive to woodfuel scarcities will likely miss the point. Even an abundance of trees may obscure more fundamental tree-related resource constraints.

Tree planting and management interventions must be more responsive to the much broader range of the needs of the farming system. Some aid agencies, governments, NGOs, and local organizations have of course done better than others in adopting a broader view of trees within development projects. The experience of programmes which have encouraged tree growing to provide multiple outputs has reinforced the assessment that farmers widely value trees for a variety of inputs into their household and farming systems, and will pursue tree growing strategies which provide as large an aggregate as possible of multiple benefits (see, for instance, Chapters 3 and 6, this volume).

The obvious point is that forestry programme planners must have an idea of what the farmer's tree-based constraints are before responsive interventions can be developed. It is counterproductive to undertake a baseline survey during the design of a 'woodfuel programme' because there is the presupposition (no doubt reinforced by a yawning 'woodfuel gap') that woodfuel is the problem. It would indeed be an institutional dilemma if an agency which had been given the mandate to develop a woodfuel project intervention discovered that in fact woodfuel was not a serious constraint in the proposed project area.

It is becoming a much more critical feature of successful programme design that other less value-laden, but more comprehensive, exercises are undertaken, such as rapid rural appraisals, participatory programme design, agroecosystem analysis, baseline surveys, and diagnosis and design studies. Particularly where issues such as land and tree tenure and usufruct complicate the dimensions of project design, these approaches are crucial. The problem is that many aid agencies (or forest departments) are ill-equipped to carry out these types of exercises without significant training and retraining.

The 'woodfuel crisis' served a valuable and useful function as a political tool which successfully raised the awareness of development planners about the interdependencies between trees and people in developing economies. The issues, however, are much farther reaching. Trees are fundamental for truly sustainable development, but for a range of reasons. The challenge is really to find ways of responding to this broader range of household needs for tree products: for building timber, fodder, fruit, fibre, soil conservation and improvement, shade and enjoyment, and income generation, as well as for fuel.

Notes

1. This chapter is revised from an earlier article (Dewees 1989).
2. See also the analysis by de Montalembert and Clement (1983).
3. Eckholm *et al*. (1984), as well as many other studies, reach similar conclusions.

4. Consumption of oil in North America and Western Europe in 1984, totalling 2194.3 million tons, was less than it was in 1973, when consumption totalled 2330.1 million tons (Jenkins 1986).
5. A range of other variables influence the price of charcoal as well. In Sudan, these have included changes in seasonal demands and supplies which have been influenced by the price of alternative fuels, prices for agricultural crops (which influence the rate of land clearance), pressures on capital markets, efforts at price control, collusion among producers and dealers, and competing demands for charcoal in export markets. In any event, on average, real prices have shown little change over time. The most dramatic price changes have been seasonal ones, and have not been sustained over the long term.
6. As Garret (1970) points out, however, 'Decomposing plant residues have many and diverse effects upon soil and the plants growing therein, but in general the beneficial greatly outweigh the harmful effects of these decompositions on crop growth.'
7. The economic and religious role of the sacred cow in India has been the subject of much intense debate. An interesting literature summarizes some of the features of dung use, cooking, and the sacred cow (Brown 1957; Heston 1971; Diener *et al.* 1978; Harris 1978; Simoons 1979; Srinivasan 1979; Lodrick 1981). Fieldhouse (1986) provides a fine overview of the anthropology of food.
8. Little information is available about the productivity of dry woodland forests when they are regularly coppiced or pollarded. Most analysts of the woodfuel situation fail to point out that woodland productivity could perhaps be greatly increased as a result of regular harvests.
9. These field surveys, reported in du Toit *et al.* (1984), were part of a Baseline Study carried out for the Zimbabwe Forestry Commission's Rural Afforestation Project.
10. See for instance, Messer (1984).

References

Agarwal, B. (1986). *Cold hearths and barren slopes*. Zed Books, London.
Anderson, D. and Fishwick, R. (1984). *Fuelwood consumption and deforestation in African countries*, Staff Working Paper No. 704. World Bank, Washington, DC.
Bajracharya, D. (1983). Deforestation in the food/fuel context: historical and political perspectives from Nepal. *Mountain Research and Development*, **3**, 227–40.
Barnard, G. and Kristoferson, L. (1985). *Agricultural residues as fuel in the Third World*. Earthscan, London.
Benians, G.J and Barber, D.A. (1974). The uptake of phosphate by barley plants from soil under aseptic and non-sterile conditions. *Soil Biology and Biochemistry*, **6**, 195–200.
Blaikie, P. and Brookfield, H. (1987). *Land degradation and society*. Methuen, London.
Brown, W.N. (1957). The sanctity of the cow in Hinduism. *Madras University Journal*, **28**, 29–49.
Brumfiel, E.M. (1980). Specialization, market exchange, and the Aztec state: a view from Huexotla. *Current Anthropology*, **21**, 459–78.
Campbell, B.M. (1987). The use of wild fruits in Zimbabwe. *Economic Botany*, **41**, 375–85.
Campbell, B.M. and du Toit, R.F. (1988). Relationships between wood resources and

use of species in the Communal Lands of Zimbabwe. *Monographs in Systematic Botany of the Missouri Botanical Garden*, **25**, 331–41.

Cecelski, E. (1987). Energy and rural women's work: crisis, response and policy alternatives. *International Labour Review*, **126**, 41–64.

Collins, J.L. (1983). Seasonal migration as a cultural response to energy scarcity at high altitude. *Current Anthropology*, **24**, 103–4.

Council on Environmental Quality and the US Department of State (1982). *The Global 2000 report to the President*, Vol. I. Penguin, London.

Dewees, P.A. (1983). Wood balance and market studies: a preliminary assessment for the World Bank Gujarat Community Forestry Project. Unpublished report. International Institute for Environment and Development, Washington, DC.

Dewees, P.A. (1985). Commercial fuelwood and charcoal production, marketing, and pricing in Kenya. Unpublished report. Joint UNDP/World Bank Energy Sector Management Assistance Programme, Washington, DC.

Dewees, P.A. (1987). Charcoal and gum arabic markets and market dynamics in Sudan. Unpublished report. Joint UNDP/World Bank Energy Sector Management Assistance Programme, Washington, DC.

Dewees, P.A. (1989). The woodfuel crisis reconsidered: observations on the dynamics of abundance and scarcity. *World Development*, **17**, 1159–72.

Diener, P., Nonini, D. and Robkin, E.E. (1978). The dialectics of the sacred cow: ecological adaptation vs. political appropriation in the origins of India's cattle complex. *Dialectical Anthropology*, **3**, 221–41.

Dirks, R. (1980). Social responses during severe food shortages and famine. *Current Anthropology*, **21**, 21–44.

Eckholm, E., Foley, G., Barnard, G. and Timberlake, L. (1984). *Fuelwood: the energy crisis that won't go away*. Earthscan, London.

Faki Ali, G. el (1985). *Charcoal marketing and production economics in Blue Nile*. National Council for Research (Energy Research Council), Khartoum.

FAO (1981). *The fuelwood situation in the developing countries*. Map prepared by the Forestry Department, Food and Agriculture Association of the United Nations, Rome.

FAO (1986). *Tree growing by rural people*, Forestry Paper No. 64. FAO, Rome.

Faulkingham, R. (1978). Where the lifeboat ethic breaks down. *Human Nature*, **1**, 32–9.

Fieldhouse, P. (1986). *Food and nutrition: customs and culture*. Croom Helm, London.

Fisher, A.C. (1979). Measures of natural resource scarcity. In *Scarcity and growth reconsidered* (ed. V.K. Smith). Johns Hopkins University Press, Baltimore.

Foley, G. (1987). Exaggerating the Sahelian woodfuel problem? Unpublished report. Panos Institute, London.

Foley, G., Moss, P. and Timberlake, L. (1984). *Stoves and trees*. Earthscan, London.

French, D. (1986). Confronting an unsolvable problem: deforestation in Malawi. *World Development*, **14**, 531–40.

Garret, S.D. (1970). *Pathogenic root-infecting fungi*. Cambridge University Press, Cambridge.

Harris, M. (1978). India's sacred cow. *Human Nature*, **1**, 28–36.

Heston, A. (1971). An approach to the sacred cow of India. *Current Anthropology*, **12**, 191–209.

Howe, C.W. (1979). *Natural resource economics*. John Wiley and Sons, New York.

Hyman, E.L. (1982). Loan financing of smallholder treefarming for woodfuel production in the province of Ilocos Norte, Philippines. Unpublished report. East-West Centre, Honolulu.

Jenkins, G. (1986). *Oil economist's handbook*. Elsevier Applied Science, London.

Klee, G.A. (1980). *World systems of traditional resource management*. Edward Arnold, London.

Leach, G. (1987). *Household energy in South Asia*. Elsevier Applied Science, London.

Lewis, O. and Barnouw, V. (1958). *Village life in northern India*. University of Illinois Press, Urbana.

Lodrick, D.C. (1981). *Sacred cows, sacred places*. University of California Press, Berkeley.

Loutit, M.W. and Brooks, R.R. (1970). Rhizopore organisms and molybdenum concentration in plants. *Soil Biology and Biochemistry*, **2**, 131–5.

Lynch, J.M. and Poole, N.J. (ed.) (1979). *Microbial ecology: a conceptual approach*. Blackwell Scientific Publications, Oxford.

McGregor, J. (1991). Woodland resources: ecology, policy and ideology. Unpublished PhD thesis. Loughborough University of Technology.

Messer, E. (1984). Anthropological perspectives on diet. *Annual Review of Anthropology*, **13**, 205–49.

Molnar, A. (1981). Economic strategies and ecological constraints: The case of the Khan Magar of north west Nepal. In *Asian Highland Societies* (ed. C.von Furer Haimendorf). Sterling, New Delhi.

Montalembert M.R. de and Clement, J. (1983). *Fuelwood supplies in the developing countries*, Forestry Paper No. 42. FAO, Rome.

Nag, M., White, B.N.F., and Peet, R.C. (1978). An anthropological approach to the study of the economic value of children in Java and Nepal. *Current Anthropology*, **19**, 293–306.

O'Keefe, P., Raskin, P. and Bernow, P. (ed.) (1984). *Energy and development in Kenya: opportunities and constraints*. Beijer Institute and the Scandinavian Institute of African Studies, Stockholm and Uppsala.

Pant, S.D. (1935). *The social economy of the Himalayas*. George Allen and Unwin, London.

Park, M. (1799). *Travels into the interior of Africa*. William Bulmer and Company, London.

Pathak, B.S. (1980). Energy balances and the use of agricultural waste on a farm. Unpublished report. Punjab Agricultural University, Ludhiana (India).

Raskin, P. (1986). *LEAP: A description of the LDC energy alternatives planning system*. Beijer Institute and the Scandinavian Institute of African Studies, Stockholm and Uppsala.

Rocheleau, D. (1985). *Criteria for re-appraisal and re-design: intra-household and between-household aspects of FSRE in three Kenyan agroforestry projects*, Working Paper No. 37. International Council for Research in Agroforestry, Nairobi.

Schneider, D.M. (1957). Typhoons on Yap. *Human Organization*, **16**, 10–15.

Scoones, I. (1989). Patch use by cattle in a dryland environment: farmer knowledge and ecological theory. In *People, land and livestock*, Proceedings of a workshop on socio-economic dimensions of livestock production in the communal lands of Zimbabwe, 12–14 September 1988, (ed. B.Cousins), pp. 277–310. Centre for Applied Social Sciences, University of Zimbabwe, Harare.

Shiva, V., Sharatchandra, H.C. and Bandyapadhyag, J. (1982). Social forestry: no solution in the market. *The Ecologist*, **12**, 158–68.

Simoons, F.J. (1974). The purificatory role of 'the five products of the cow' in Hinduism. *Ecology of Food and Nutrition*, **3**, 21–34.

Simoons, F.J. (1979). Questions in the sacred cow controversy. *Current Anthropology*, **20**, 467–93.

Singer, S.F. (1984). World demand for oil. In *The resourceful earth: a response to Global 2000* (ed. J.L.Simon and H.Kahn). Basil Blackwell, Oxford.

Sownmi, M.A. (1985). The beginning of agriculture in West Africa: botanical evidence. *Current Anthropology*, **26**, 127–9.

Srinivasan, D. (1979). *Concept of cow in the Rig Veda*. Motilal Barsidass, Delhi.

Swift, J. (1977). Sahelian pastoralists: underdevelopment, desertification, and famine. *Annual Review of Anthropology*, **6**, 457–78.

Toit, R.F. du, Campbell, B.M., Haney, R.A. and Dore, D. (1984). *Wood usage and tree planting in Zimbabwe's Communal Lands*. Forestry Commission of Zimbabwe and the World Bank, Harare.

Weatherly, W.P. (1980). Environmental assessment of the Rural Electrification I Project in Indonesia. Unpublished report. Agency for International Development, Washington, DC.

Wiersum, K.F. and Veer, C.P. (1983). Loan financing of smallholder treefarming in Ilocos, a comment. *Agroforestry Systems*, **1**, 361–5.

Wilson, K.B. (1989). Trees in fields in Southern Zimbabwe. *Journal of Southern African Studies*, **15**, 369–83.

Wu, W. and Chen, E. (1983). Our views on the resolution of China's rural energy requirements. *Biomass*, **3**, 287.

8 Wood product markets as incentives for farmer tree growing

Peter A. Dewees
N.C. Saxena[1]

8.1 Trees, markets, and agricultural policies

In order better to understand how markets for tree products have played a role in encouraging households to incorporate trees into their farming systems, a broader view of the operation of agrarian economies is useful. The potential impact of policy interventions and project innovations can be better understood within a framework which addresses the number of different markets in which rural households participate. Policies toward food production, in particular, offer a number of useful paradigms in exploring how farmers respond to different market signals.

Farmers typically operate at the intersection of three groups of markets:*

- markets for agricultural commodities, the sale of which determines the household's revenues;
- markets for factors of production (physical inputs as well as land, labour, and capital), which, in conjunction with revenues earned from the sale of agricultural commodities, determine the household's level of income; and
- markets for consumer goods, particularly for commodities manufactured in cities. The prices which rural households pay for these commodities tend to determine the real value of farm income.

The ways in which these markets interact, and particularly, the ways in which agricultural prices and policies influence these interactions, play a key role in agrarian production.

Certainly with regard to trees on farms, rural households stand at a similar conjunction of markets. This chapter discusses how markets for tree-based goods have influenced the decision by farmers to use their land for planting trees. Most analyses of the relationship between markets and trees have limited their view of the issue to this single perspective. The chapter which follows this one, however, explores the relationship between tree growing and

---

* This discussion is in part derived from Bates (1981).

other markets, particularly those which help to define household land, labour and capital allocation processes.

In some respects, where markets for wood and tree products have developed, the analysis is far more straightforward than the problem of understanding how the production of tree-based goods, used for subsistence purposes by households which may operate outside of land, capital and labour markets have provided incentives for tree growing. Prices and costs of production are far better understood where there are markets to be exploited. Conceptualizing the impact of market supply and demand on tree growing is far easier than establishing how subsistence demands may influence farmers' land-use decisions.

Even so, tapping into markets, and into the process of specialization and exchange, in order to encourage rural households to incorporate trees into their farming system has proven to be a challenge for many rural forestry project planners. Where markets have turned out to provide significant incentives for tree growing, this has usually been understood only after-the-fact. It has been more commonly the case that forestry planners have recognized the importance of markets *after* project investments have been made, usually in tree planting options which were intended for vastly different purposes.

This chapter is organized around a discussion of three particular groups of markets for wood-based products: the charcoal market in Khartoum; markets for products from black wattle woodlots in Kenya; and markets for products from eucalypt woodlots in India. The first case discusses the operation of an extremely sophisticated market for charcoal, and why the introduction of sustainable systems of woodfuel growing would be highly problematic. The second case describes a system of sustainable production which developed in response to market signals. The third case describes a number of interventions which led to the widespread adoption of tree planting practices, only to result in a glut of timber on rural and urban markets, and a collapse in the market. Finally, the chapter closes with a discussion of some of the implications these cases pose for policy and planning processes.

8.2 Urban charcoal markets in the Sudan

Introduction

The charcoal market in Sudan is characterized by a complex network of entrepreneurs, agents, and labourers who produce huge quantities of charcoal in highly-efficient earth kilns, operating from as far away as 500 km from urban markets. This type of production is dependent on large, contiguous areas of savannah forests which, for the most part, are being cleared for agricultural development.[2] The production of roundwood in sustainable management systems for conversion into charcoal is not a feature of the market. One of the problems inherent in introducing new systems of production which would be dependent on these types of sustainable systems of management, would be that such systems would have to compete in a well established, highly competitive, and extremely sophisticated business environment.

It was estimated in 1985 that Blue Nile and Kassala Provinces supplied as much as 80 per cent of the Khartoum area's demands for charcoal. Eighty-five per cent of production in these areas came from agricultural land clearing schemes.[3] Production is controlled, financed, and organized by around 25 of the leading businessmen in the trade. These, in turn, employ between one and three foremen/agents who oversee production at particular sites and are responsible for recruiting anywhere from 50 to 100 labourers to work at each site. Labourers are drawn from a large pool of seasonal, migrant workers, mostly from western Sudan, who are employed primarily in mechanized rain-fed farming schemes.

The charcoal production cycle

Charcoal production activities are financed by credit arrangements which are made between businessmen, agents, and labourers, and are to a great extent dependent on the seasonal storage and sale of charcoal to generate working capital.* Agents travel to western Sudan in September or October to recruit charcoal labourers. Labourers are provided with an advance to travel to main production centres. During the earliest period of labour recruitment, charcoal prices are highest, and working capital is often generated by the sale of charcoal which has been stored since the previous season. This helps to finance further labour recruitment as the charcoal burning season gets under way.

As soon as charcoal production has commenced, additional capital is generated through the sale of charcoal, more labour is hired, and production accelerates. Some production will be stored in anticipation of higher demands between July and January and in anticipation of capital requirements to recruit labour in the fall, but the balance will be sold to settle immediate labour accounts. The combination of lower market demands for charcoal, and the need for greater liquidity to settle labour accounts during this period causes prices to gradually drop, bottoming out in April and May. Production slows down in June and ceases completely from July. The sale of charcoal stored in depots begins later in the year to finance the recruitment of labour in September, for burning during the following season, which begins again around November.

The charcoal business is heavily dependent on this system of revolving credit. As production is sold, returns are reinjected, and it becomes largely self-financing by allowing initial investments of working capital to revolve

* This discussion is largely derived from el Faki Ali (1985) and el Nour and Satie (1984).

Table 8.1 Seasonal labour and capital requirements for large-scale charcoal production in Sudan

September to October	• labour recruitment, Western Sudan; • cash advances provided to labourers as required;
November	• application for a concession; • concession granted; • cash advances provided to labourers as required; • begin burning in small kilns;
February to March	• continue burning in small kilns; • sell first production to take advantage of highest charcoal prices and to generate working capital; • reinvestment in additional labour; • expand production as labour becomes available;
April to May	• accelerate production schedule to levels of peak production; • labour bills accumulate; • liquidity requirements increase; • charcoal sales accelerate to generate liquidity so investor can settle labour accounts; • high production and need for liquidity causes drop to lowest charcoal prices; • charcoal storage in depots begins • production season winds down;
June	• production season ends; • final labour bills are settled; • production in excess of liquidity requirements is stored in depots;
July to November	• rainy season; • charcoal production ceases completely;
September to October	• labour recruitment for the next season commences.

Sources: el Faki Ali (1985) and el Nour and Satie (1984).

through the system. The annual cycle of charcoal production activities is summarized in Table 8.1.

Pricing structure and the costs of production

The principal source of wood used for making charcoal comes from the clearing of large areas of Acacia woodlands to support an expansion of rain-fed mechanized agriculture. Since the early 1950s, enormous areas of these forests and woodlands have been cleared, mostly for sorghum production.

The cost of wood used for making charcoal, accounted for by royalties collected by the Forestry Administration and by local authorities, makes up a very small percentage (around 4 per cent) of the Khartoum retail charcoal price. Despite the low cost of wood and the potential profitability of charcoal production, huge areas of woodlands have been cleared and burnt on-site without conversion into charcoal. It has been estimated that between 1979 and 1984, in just two mechanized farming sites in Blue Nile Province, nearly 700 000 *feddans* were cleared and burnt on-site resulting in the loss of nearly 21 million sacks of charcoal.*

The market framework becomes quite complex from the point of production or storage to the market. The more integrated an entrepreneur is in the market structure, the greater the profit margin and the greater the ability to compete in the market. For instance, if an entrepreneur controls production, transport, and distribution, losses in one sector which might be incurred in order to gain a particular market share, can be offset by profits in another sector.

Charcoal transporters travelling by road play a key role in ensuring that supplies reach urban markets in a timely and efficient manner. Charcoal transport can be especially cheap, as it is seen as a means of carrying a profitable load to Khartoum or to Port Sudan in order to pick up more valuable primary commodities for transport to the interior. Transport distances from centres of production in Blue Nile Province to Khartoum are around 470 km.

Transportation costs account for between 20 and 25 per cent of the Khartoum retail price for charcoal: the largest single cost-component of the retail price. Despite the very long transport distances which are involved, by comparison with other African urban charcoal markets, these costs are low. In Kenya, for instance, transport costs account for *45 per cent* of the retail price.

In urban areas, charcoal is generally delivered to large- or medium-sized depots (*zaribas*) for both wholesale and retail selling. Wholesalers generally sell by the sack. Retailers sell by the sack and by small quantities (by volume). Retail prices of charcoal sold in small quantities may be 30 per cent higher than retail prices of charcoal sold by the sack, even accounting for losses in the sack. A breakdown of Khartoum retail prices, by price component, is given in Table 8.2.

Charcoal pricing trends

Particularly when there are longer term considerations about the likely impact of market-oriented tree planting initiatives, an understanding of how and why market prices have fluctuated in the past, and are likely to fluctuate in the

* A *feddan* is equal to 0.42 ha.

Table 8.2 Cost components of retail charcoal prices (Khartoum) for charcoal produced in Blue Nile Province

Cost item	Price per sack (S£)	Per cent of Retail Price
Production costs		
Labour	4.03	20
Supervision		
Foreman	0.45	2
Agent	0.28	1
Burial	0.19	1
Water	0.59	3
Fees, royalties, and taxes	0.80	4
Sacks	1.49	7
Production costs sub total		
Without sacks	6.34	32
Packed in sacks	7.83	39
Marketing costs/profit margins, ex-site		
Without sacks	0.68	3
Packed in sacks	1.19	6
Selling price, ex-site		
Without sacks	7.01	35
Packed in sacks	9.01	45
Depot costs (packed in sacks)		
Transport to the depot	1.17	6
Storage tax	0.16	1
Depot costs sub total	1.33	7
Marketing costs/profit margins, ex-depot	1.46	7
Selling price, ex-depot	10.61	53
Transport and distribution costs		
Guarding at the depot	1.68	8
Handling and packing	0.79	4
Losses and depreciation	0.19	1
Transport to Khartoum	4.80	24
Transport and distribution costs sub total	7.46	37
Production, depot, transport and distribution costs sub total (price, ex-truck)	16.61	83

Table 8.2 (*cont.*)

Cost item	Price per sack (S£)	Per cent of Retail Price
Marketing costs/profit margins, Khartoum		
Wholesaler I	0.13	1
Wholesaler II	1.15	6
Retail	2.11	11
Selling price, Khartoum		
Wholesaler I	16.75	84
Wholesaler II	17.89	89
Retail	20.00	100

Source: el Faki Ali (1985) updated in Dewees (1986).

future, is essential. From the perspective of both the public and private sectors, longer term pricing trends can be particularly enlightening.

Historical perspectives   Charcoal is essentially an uncontrolled commodity, and, with a few exceptions, prices fluctuate freely. Attempts at controlling prices have had little long-term impact. Such efforts have been unenforceable, have contributed to shortages, and have quickly collapsed.

While nominal prices for charcoal have increased quite substantially over 10 years from 1976 to 1986 (nearly 15–fold), except for quite dramatic short term and seasonal fluctuations *real prices have shown little long-term change*. Figure 8.1 indicates the extent to which real Khartoum retail prices have changed over time.

Periodic fluctuations in the price of charcoal are the result of a range of factors, and are not solely accounted for by an abundance or scarcity of roundwood for burning into charcoal. Indeed, as the price of roundwood accounts for such a small proportion of the retail price, it would be difficult to conclude that changes in the price of charcoal are much of a reflection of the physical scarcity of woodfuel. Price fluctuations are likely accounted for by other factors. For instance:

- *Petroleum price increases*, particularly in 1978 and in 1981–1982, increased the cost of diesel fuel for the transport sector, and of bottled gas for household use. Higher prices for transport, and a greater reliance on charcoal as a household cooking fuel could have brought about higher charcoal prices during these periods.
- *Good agricultural harvests* could have contributed to higher demands for agricultural labour, resulting in higher labour costs for charcoal entrepreneurs, and higher costs for urban charcoal consumers.

- *High prices for agricultural crops* could have contributed to higher rates of agricultural land clearing, meaning that there would be more wood available for burning into charcoal, a greater supply of charcoal on the market, and lower prices for urban charcoal consumers.
- *Pressures on capital markets* which have traditionally financed charcoal entrepreneurs could have limited their access to start-up capital, resulting in higher charcoal prices.
- *Efforts at price control* in urban areas could have contributed to hoarding and to the refusal of producers, transporters, wholesalers, or retailers to sell on the open market. Resulting scarcities could have caused higher prices.
- *Collusion among producers and dealers* could have forced the price up over short periods. The need for quick liquidity and the revolving working capital concept, however, all work against collusive behaviour.
- *Charcoal exports* to the Middle East, although not currently a feature of the market, have in the past caused scarcities and higher prices.

For the most part data sets which could be used to test these hypotheses are incomplete or are otherwise unsatisfactory.

Future pricing trends   Because charcoal is generally produced as a salvage operation, as a by-product of agricultural land clearing, the availability of future supplies will be defined by the extent of future land clearing activities. Large-scale clearance for mechanized farming is expected to continue, but at

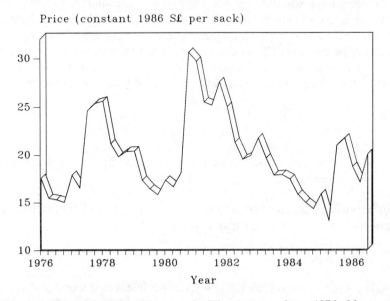

Fig. 8.1 Charcoal retail pricing trends, Khartoum markets, 1976–86.

a slower rate. Mechanized farming is expected to move into South Kordofan, Upper Nile, and White Nile Provinces, and into Southern Darfur, and so these areas are likely to become major sources of charcoal in the future.

The two most significant costs of the delivered price of charcoal in Khartoum are labour and transportation costs, accounting for 20 and 24 per cent, respectively, of the retail price of charcoal. Labour costs in new production areas are likely to be lower, though transportation costs will be much higher. It has been estimated that, assuming all other costs of production remain constant, the net effect of shifting centres of charcoal production from Blue Nile Province to areas away from Khartoum, but where labour is cheaper, would be to increase the retail price by around 25 per cent. Assuming that these changes in the structure of production are made over 5 years, prices could be expected to increase at a rate of around 5 per cent annually.

The impact of these changes in the structure of charcoal pricing would be to increase the share of transportation costs and to decrease the share of labour costs as a per centage of the retail Khartoum price. Labour costs, for instance would decrease from 20 to 11 per cent of the retail market price of charcoal. Transportation costs would increase from 24 to 49 per cent of the retail market price of charcoal, assuming all other cost shares and profit margins stay constant.

Developing sustainable systems of production to meet urban market demands

In determining whether or not sustainable production systems could be developed to compete in urban charcoal markets in Sudan, the delivered price of charcoal from two potentially competing sources must be compared. The ability of multiple sources of charcoal in Sudan to compete in the same market is a function of the cost of wood at the source, the cost of burning charcoal and loading it on to transport, and the cost of transport to the market, or:

(1) $P = CW + CC + TR$
where P = the price in the market off-the-truck;
CW = the cost of wood;
CC = the cost of burning charcoal and loading it on to transport; and
TR = the cost of transport to the market.

For a source of charcoal derived from B to compete with a source of charcoal derived from A, it must be true that $P_A \geqslant P_B$.

(2) $P_A = CW_A + CC_A + TR_A$;
(3) $P_B = CW_B + CC_B + TR_B$.

If $CW_A + CC_A$ is less than $CW_B + CC_B$ (for instance if wood at A is acquired at no cost from natural forest clearing and wood from B is grown in woodlots

Table 8.3 Current radii of haulage within which sustainably produced charcoal could compete in urban markets in Sudan

Distance between markets and current supplies (km)	Urban market				
	Khartoum	Wad Medani	El Obeid	Kassala	Port Sudan
	470	210	300	100	450
Stumpage value of wood (in S£/m³)	Radius of haulage in kilometres, within which savings in transport costs offset the cost of sustainable production of wood for conversion into charcoal				
5.0	420	182	263	83	401
10.0	194	65	107	21	183
15.0	71	15	31	2	66
20.0	17	1	4	–	15
25.0	1	–	0	–	1

Notes: See Annex to this Chapter. – indicates it would be cheaper to continue transporting charcoal from current sources of production. Charcoal conversion efficiency for both sustainable and non-sustainable production is assumed to be 29 per cent. Charcoal is transported in 20-ton lorries. The density of native hardwoods used for charcoal is estimated to be 0.828 tons per solid cubic metre. The density of sustainably grown wood for charcoal, such as *E camaldulensis*, is assumed to be 0.7 tons per solid cubic metre. Royalties and fees for traditional charcoal producers are estimated to total S£ 0.80 per sack of charcoal, or around S£ 4.9 per cubic metre of roundwood. Other costs of charcoal production, including costs at the site, depot costs, and distribution costs, but excluding transport costs and wholesale and retail profit margins total S£ 11.02 per sack of charcoal.
Source: Dewees (1986).

or plantations, and charcoal production costs are constant), then in order for charcoal derived from B to be competitive in the market, TR_A must be greater than TR_B by the difference between the costs of charcoal and wood production to the truck at A and B. In other words, if charcoal can be produced from woodlots or plantations closer to the market at B, it may have some advantages over charcoal produced at A if it can be produced for savings in transport costs which would otherwise be spent travelling the greater distance to A instead of to B.

By substituting data collected from a series of empirical studies into these equations (which are more fully developed in the Annex to this Chapter), the financial viability of growing wood for conversion into charcoal can be evaluated. The results of this analysis are shown in Table 8.3, which reflects the market in 1986 when empirical data were collected about average transportation distances and costs. A radius of haulage can be defined around each urban market within which savings in transportation costs (which would otherwise be incurred by relying on sources of production taken from greater distances) offset the cost of sustainably producing wood for charcoal conversion.

Stumpage rates in Table 8.3 reflect the range of costs of producing roundwood on-the-stump. Particularly in light of the very slow growth of roundwood, and the high costs of plantations (which are usually irrigated), it is unlikely that roundwood could be produced from dedicated woodfuel plantations for much less than S£30 /m³ (in 1986 prices).

But at existing transportation costs and radii, if it costs much more than S£25 to produce a cubic metre of wood, transportation cost savings alone will not make sustainable charcoal production systems viable. Even at lower stumpage rates, it is going to be cheaper to transport traditionally produced charcoal from great distances, except within a radius very close to urban markets.[4]

Transportation distances and costs *will* increase as standing stocks are depleted, and as the boundaries of agricultural development are pushed further out. To reflect this, economic radii of haulage can be recalculated if increased transportation costs are considered. These are shown in Table 8.4.

Table 8.4 Future viability of sustainably-produced charcoal in Sudan

Estimated distance between markets and future supplies (km)	Urban market				
	Khartoum	Wad Medani	El Obeid	Kassala	Port Sudan
	757	338	483	161	725
Stumpage value of wood (in S£/m³	Radius of haulage in kilometres, within which savings in transport costs offset the cost of sustainable production of wood for conversion into charcoal				
5.0	686	298	432	138	656
10.0	357	126	201	44	338
15.0	158	39	75	8	148
20.0	54	6	18	0	49
25.0	11	0	1	–	9

Notes: See notes in Table 8.3 and discussion in text. Reflects the likely economic radius of haulage for sustainably produced charcoal, assuming that transportation costs increase by 10 per cent/year over the next 5 years. Future estimated radius assumes transportation costs per ton-km remain constant.
Source: Dewees (1986).

The basic point remains the same as that made in Table 8.3. *Except for wood which is grown very cheaply and close to urban markets, it will be cheaper to continue transporting traditionally produced charcoal from great distances for some time to come.* Under these assumptions, there is little likelihood that private sector or unsubsidized public sector wood production will end up on the charcoal market.

Summary

This example shows how difficult it can be to grow woodfuel in a way which is financially viable, even when there is a seemingly lucrative urban market. Planners often assume that, simply because woodfuels are being transported from very considerable distances, it would be relatively straightforward and financially viable to grow wood closer to an urban market. Comparative advantage in these cases must be clearly thought through: what price advantage will systems of sustainable production have over existing sources of supplies and markets?

The following case explores the market conditions for tree products which prevailed in Kenya, from the 1920s through the early 1960s, and which encouraged a large number of small farmers to plant trees. Kenyan farmers were quite able competitively to grow tree products to meet particular market demands for a combination of reasons having to do with the low value of alternative land uses, land tenure insecurity, and labour supply problems.

8.3 Black wattle woodlots in Kenya

The development of markets for wattle bark

Black wattle (*Acacia mearnsii*) was probably introduced into Kenya sometime in the 1890s. Although indigenous to Australia, the seed used for Kenyan plantings was likely brought by settlers from South Africa, where it had been widely planted to provide timber for the mining industry. The bark of black wattle is particularly high in tannins. European farmers in Kenya began planting it so the bark could be harvested and sold to the tanning extract industry from around 1915; by 1919, around 4000 ha of plantations had been established on European farms.

African farmers were first encouraged to plant wattle from around 1911. The Colonial Government's original objectives were to reduce pressures on indigenous forests, which had been closed to further agricultural expansion and settlement. A push by the Government to encourage the planting of trees on farms began in earnest around 1921, but was widely resisted.

The situation radically changed by the late 1920s, however, as wattle came to be widely adopted, particularly among Kikuyu farmers in Murang'a, Nyeri, and Kiambu Districts, as a response to the bark trade.

In Kikuyu, small black wattle plantations are numerous and in more accessible areas, their increase has been encouraged by the trade in wattle bark. (Colony and Protectorate of Kenya 1928)

There is no organized planting, but the planting of black wattle is extending rapidly and thousands of small wattle clumps are quite rapidly changing the appearance of many parts of the Reserve. (Colony and Protectorate of Kenya 1930)

Within a few years, the Forest Department claimed that 'the afforestation problem had been solved' in Kikuyu Province (Colony and Protectorate of Kenya 1932).[5]

There was limited capacity to process the bark at this early stage in the development of the industry, and most of it was exported. Tanning extract factories were opened in the early 1930s. Increased access to markets for bark, which accompanied the establishment of these factories, greatly contributed to the popularity of wattle as a smallholder crop. In a single year, 1935, it was reported that the total area under wattle had nearly doubled since the previous year, from 18 000 to 40 000 ha (Colony and Protectorate of Kenya 1935; Kitching 1980). By 1937, it was estimated that there were nearly 18 000 ha of wattle in Murang'a District alone, compared with 9700 and 6100 ha in Nyeri and Kiambu Districts, respectively.

Although the earliest planting campaigns were directed at a relatively small class of the educated elite, later programmes were much further reaching and introduced wattle to the widest possible range of farmers. Wattle production came to be concentrated on the holdings of the middle peasantry, primarily because it produced a broad range of household commodities (fuelwood, charcoal, building poles, wood for the construction of farm buildings and cattle enclosures, and so on) while, at the same time, generating household income (Cowen 1978).

Wattle production could be especially profitable. Africans were, for the most part, excluded from the production of other cash crops, and returns to food crops such as maize and potatoes were, by comparison, quite low. In 1934, the Nyeri District Commissioner J.W. Pease noted in the report of the Kenya Land Commission that

. . . at the moment, the price of a ton of chopped wattle bark is about the same as the price of a ton of native maize. If this continues I shall anticipate a much greater increase than 50 per cent in land under wattle in this district; it might easily be doubled in the next 10 years. (Government of the United Kingdom 1934)

The situation remained very much the same until the late 1950s. Until land reform and Independence in the early 1960s, wattle was really the only cash crop which Africans were encouraged to cultivate. In this respect it played a major role in income generation, and in the political processes which accompanied social differentiation.

Wattle continues to be planted. A third of wattle woodlot-operating households interviewed in Murang'a District, for instance, in 1990 reported that they had established their woodlots since 1980 (Dewees 1991). Wattle covers around 6000 ha of good quality farmland in Murang'a, compared with 10 000 ha under tea, and around 40 000 ha under coffee. For the most part, wattle is extensively managed, with few labour or capital inputs, and many farmers see it as a relatively productive way of using fallow land.

Land tenure and wattle woodlots

Until the mid-1960s, wattle bark production was indeed lucrative, and relatively easy *vis-à-vis* other land-use options. But wattle was also popular because of its implications for land tenure. Land tenure for Africans during the 1920s and 1930s was especially precarious. Forests had been closed to agricultural expansion, large areas of land were expropriated by white settlers, and Kikuyu farmers, for the first time, began to find that there was simply not enough land to guarantee that everyone had a share. The development of the wage labour market encouraged African men to live and work off the farm. If they were unable to keep their holding under cultivation, land-use rights in the Reserves would be lost. Households not engaged in farming tended to lend cultivation rights to others as a way of keeping their land under cultivation, and hence, of retaining land-use rights.

Customary land borrowing and lending arrangements were hereditary. After a number of generations though, the tenure picture in Kikuyu areas had become especially confused, and claims and counterclaims on the same plots of land were common. So, in addition to its income-generating potential, wattle was favoured because it meant that men could retain customary land-use rights, asserted by cultivation, by planting wattle, thus enabling them to seek wage employment elsewhere without having to lend the land to someone else and to jeopardize their own land-use rights.

The land tenure situation encouraged farmers to plant wattle for a number of other reasons as well. Land claims could only be settled by the courts. The process of litigation was costly. Wattle generated income, which enabled households to make their claims in the courts. Finally, particularly in light of the confusing tenure situation which had prevailed for some time, wattle helped to *establish* rights to land which could be greatly strengthened by the establishment of permanent crops, particularly if borrowed land had been cultivated by tenant families for several generations.

Woodfuel markets and wattle production

The trend toward diversifying production   With the land reforms which gathered pace in the late 1950s, and with the introduction of a range of better-paying cash crops, wattle became far less popular than it had been. Bark prices also fell (by 50 per cent between 1955 and 1962). After Independence, large areas of wattle were cleared and more lucrative cash crops, particularly coffee and tea, were planted. Rather than disappearing altogether, however, wattle came to be managed for a range of products, rather than simply for its bark. Wattle roundwood was ideal for charcoal, and could be used for fence posts and construction poles as well. Even during the peak periods of bark production, it was a valuable source of charcoal, and continues to be so.

Indeed, markets for charcoal made from wattle trees were nothing new. Wattle woodlots around Nairobi began providing significant supplies of woodfuel from the early 1930s in response to increased urban domestic and industrial demands.

. . . The demand for fuel for domestic and industrial purposes around Nairobi is increasing rapidly and is becoming increasingly difficult to satisfy owing to the limited area of forest within an economic distance of Nairobi . . . It is likely that the price of fuel will drop when wattle plantations around Nairobi mature. (Colony and Protectorate of Kenya 1929)

In 1945, between 700 and 1000 sacks of wattle charcoal *a day* were being transported the short distance from Kiambu District to Nairobi to meet urban fuel demands (Gollop 1945). Murang'a District also provided significant supplies of charcoal to meet domestic and export market demands. Figure 8.2 shows trends in the charcoal and wattle bark trade which originated in Murang'a District from 1949 through 1974.

Management strategies for bark production and for charcoal are different. The best tannins are processed from trees which are at least 7 years old. Charcoal, by comparison, can be produced from relatively small trees, as

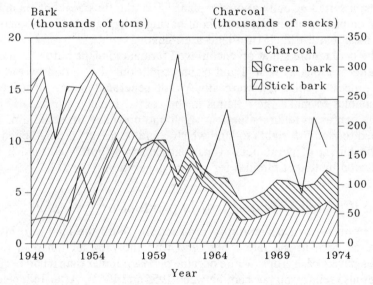

Fig. 8.2 Wattle bark and charcoal exports from Murang'a District, Kenya, 1949–74.

young as 4 years old. While the production of high quality wattle bark depends on there being a relatively low density of older trees in a woodlot, returns to charcoal production can be maximized by the management of very dense stands of trees on short rotations. As prices for tannins fell and as prices for charcoal increased, farmers tended to shorten the rotation of their woodlots in order to produce charcoal. Because these trees were too young to produce high quality bark, bark production fell.

A driving force behind charcoal production from wattle were the reforms and changes in agricultural production which preceded and immediately followed Independence. As the pace of land reform increased, large areas of wattle were cleared and planted with other crops. Charcoal was produced from the cleared woodlots, accounting for the sixfold increase in production in Murang'a between 1952 and 1961.

Charcoal production from wattle   It is characteristic of charcoal production from wattle that conversion efficiencies are high and that the quality of the charcoal is quite good. Elmer (1943) described medium-scale wattle charcoal production using traditional earth kilns in the early 1940s. He could have been describing contemporary production practices. Wattle timber is stacked, loosely packed with dry twigs and sticks, then covered with a layer of sod and sealed so it is nearly airtight. Vents are left in one end of the stack. An opening is made at the other end of the stack and fire is introduced. Once lit, the stack is closely watched, and airflows are carefully regulated to prevent overcombustion. The burning and cooling cycle takes 5–10 days, after which the charcoal is packed into maize sacks (*gunia* in swahili). A single *gunia* holds 30–35 kg of charcoal. A team of four or five experienced burners can construct and burn up to a dozen stacks over a month.

Arrangements for the payment of burners vary. Among burners interviewed in Murang'a, the most common arrangement seemed to be that the owner of the woodlot would evenly split earnings from the sale of the bark and the charcoal with the burners. Kinyanjui (1987) estimated that tree-owners earn around 60 per cent of the gross revenues produced by wattle. In any event, considering the fact that the owner's costs of production are low, consisting largely of the costs of allowing trees to naturally regenerate on underutilized fallow land, this generates very high returns for a very small investment.

As a source of urban woodfuel, wattle is an exception in the sense that charcoal producers have implicitly paid something for the roundwood. This is usually not the case, as, in Sudan, most producers pay virtually nothing for their wood. In land clearance operations, charcoal production is carried out as a salvage operation, and the cost of charcoal at the roadside reflects only the cost of labour for production. The result is that the roadside cost of charcoal produced by salvage operations comes nowhere near to reflecting either its economic or replacement value.

Wattle charcoal, on the other hand, is made from roundwood which has an

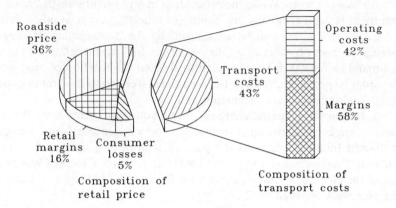

Composition of
retail price

Composition of
transport costs

The wholesale price is equal to the
roadside price plus transport costs.

Fig. 8.3 Composition of Nairobi retail price of charcoal, sold by the sack.

implicit value. Recalling that wattle tree owners recover 60 per cent of the gross revenues from charcoal production, it could be argued that this represents the real stumpage value of wattle timber, the balance being accounted for by conversion and production costs to the roadside.

In 1990, a hectare of mature wattle when harvested after 8 years could produce around KSh 3870 for its bark, and around KSh 28 810 for charcoal at the roadside (a total of around US$1100 at prevailing 1990 exchange rates). Even if a household split the roadside earnings with charcoal burners and bark strippers, net income to the household would still total over KSh 16 000 /ha: an exceptional return to the household in that it reflects capital and labour costs which were virtually nil.

Unlike in the Sudan, urban charcoal markets in Kenya show no seasonal price fluctuations. In wattle growing areas at least, there is no annual cycle of charcoal production *per se*. The market is far less vertically integrated than in the Sudan, and transporters (rather than entrepreneurs who control all aspects of the market) are the crucial link between rural producers and urban markets. In the absence of vertical integration, and with the dominance of transporters in the marketing chain, the result is, not surprisingly, that transportation costs account for a far greater proportion of the delivered market price. The composition of the Nairobi charcoal price is shown in Fig. 8.3.

Pricing trends and their impact on charcoal production

Until quite recently, charcoal prices were legally controlled, though controls tended to be selectively applied. In the early 1980s, the largest single producer of charcoal in Kenya, East African Tanning Extracts, sought to sell charcoal produced in its highly-efficient brick kilns at market prices. The District administration intervened, and told the company they would be in violation of the Price Control Act if they did so. As a result, EATEC stopped producing charcoal altogether. They continue to sell their wattle timber as *fuelwood* to anyone who wants to buy it, and make their charcoal kilns available as a 'public service.'

The price of fuelwood *is* controlled (at around KSh 43 /ton, delivered), but these price controls were last revised in 1957 (Colony and Protectorate of Kenya 1957). As a result, price controls on woodfuel were long ago forgotten, and EATEC sells its fuelwood (in 1989) for around KSh 300 /ton at the roadside (H. von Kaufmann, personal communication).

Pricing trends for charcoal and for paraffin are shown in Fig. 8.4. The use of paraffin as a household fuel has become increasingly important: between 1982 and 1988, paraffin prices fell by over 40 per cent, primarily because of subsidies, while paraffin consumption nearly doubled. Until the late 1970s, charcoal maintained significant price advantages over paraffin as a domestic fuel.

The Government's current policy towards paraffin subsidies is that they are necessary to reduce charcoal consumption. If charcoal and paraffin were indeed substitutes, it could be expected that declining prices for paraffin would be mirrored in declining prices for charcoal. Since 1985, the opposite has been

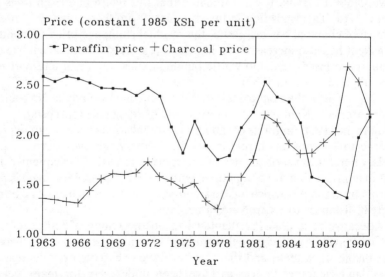

Paraffin price is in 1985 KSh per half litre;
Charcoal price is in 1985 KSh per kg.

Fig. 8.4 Charcoal and paraffin retail pricing trends, Nairobi Markets, 1963–91.

the case. Because of this, it is unlikely that heavy paraffin subsidies which have brought about low prices are currently having any impact whatsoever on rates of charcoal consumption. There is a much stronger case to be made that paraffin subsidies have served primarily to buffer consumers from price fluctuations.*

Prices and changes in the structure of the charcoal market   Until the early 1980s, improvements in the transportation infrastructure in Kenya made it possible for charcoal market prices to be moderated over the long term. More supplies became more accessible as new forest or bush areas were cleared for cultivation, and as charcoal was produced as a by-product of land clearance.

Land clearance has long provided the bulk of the charcoal used in urban areas of Kenya. The pace of settlement in new areas was sufficient to produce enough charcoal to meet urban demands. When charcoal is produced as a by-product of land clearance, labour costs of production are quite low. When land is cleared *specifically* to produce charcoal, labour costs of production increase considerably. Price increases since 1985 may reflect this change in the structure of supply, as charcoal supplies are increasingly being produced as the sole object of land clearance, rather than as a by-product of land clearance.

The other interesting development in the structure of production is the relatively larger share of charcoal on the Nairobi market which is produced as a result of the burning of wattle. In 1985, it was estimated that around 5 per cent of the market was being supplied by charcoal produced from wattle woodlots. By 1989, it was estimated that this figure had risen to as much as one-third. This transition in the structure of production makes sense. As the price of charcoal has increased, the residual stumpage price of wood used to make it has also increased. At higher prices, wattle charcoal can yield similar stumpage prices as could wattle building poles, the main alternative use for wattle timber (Bess 1989).

Most wattle charcoal comes from the immediate vicinity of Nairobi, particularly from the Kiambu and Limuru areas. Although the real costs of production are higher (both in terms of labour and roundwood costs), charcoal produced close to Nairobi has a real competitive advantage over charcoal produced elsewhere because of savings in transportation cost. Consequently, it can be sold at prices that are competitive with prices for charcoal, which is produced from bush timber (which has no stumpage cost) but which must be transported great distances to reach Nairobi markets.

Because of a 1986 Presidential ban on the felling of indigenous trees, it is conceivable that charcoal from land clearing operations has become less available since then, and that this has driven charcoal price increases. Even so, land clearance processes have been under way for many years. That they should rapidly decelerate over a few short years would be something

---

* Current policies toward paraffin subsidies are laid out in Republic of Kenya (1989).

of a surprise. Price increases of the magnitude observed since 1985 could be expected over the longer term, as the pace of land clearance slowed and as less land became available for cultivation.

Controls on charcoal production and transport, which were implemented as a result of the 1986 ban on the felling of indigenous trees, have probably had the greatest impact on urban woodfuel prices. In Murang'a District, for instance, very little wattle charcoal is produced which ends up on the Nairobi market. This is despite the fact that Murang'a is within easy proximity of Nairobi and has historically supplied large amounts of charcoal to meet urban demands.

Most charcoal production in Murang'a services rural demands in the lower zones of the district and does not pass beyond district boundaries. Transport out of the district is problematic because of movement and transport controls. In order to sell charcoal to Nairobi, charcoal producers would have to first obtain a permit from the Forest Department indicating that charcoal supplies were not obtained from the high forests, then they would have to obtain a movement permit. Permits are difficult and costly to obtain. The result is that wattle woodlots in Murang'a are underutilized for charcoal production and little ends up on Nairobi markets. Indeed, only around 3 per cent of the woodlot growing households interviewed in Murang'a in 1990 reported that they had produced and sold charcoal the last time their woodlot was harvested.

Summary

Black wattle was introduced into Kenyan farming systems long before contemporary notions of social forestry were in vogue. It was widely adopted by African farmers, mostly because of its income generating potential, but also because farmers were excluded from most other opportunities for growing cash crops. In conjunction with a precarious land tenure situation, and partly in response to emerging labour markets (discussed in Chapter 9), extensive areas of high potential agricultural land were planted with wattle. Many areas were subsequently cleared following land reform and as new land-use opportunities became possible after Independence.

Over time, markets for products from wattle became more diverse. These included markets for charcoal and for building timber. The exploitation of these markets involved different tree management approaches, such as shorter rotations and closer spacing. Even when prices for particular products collapse, alternative markets may develop.

While woodlots could sustainably produce large quantities of charcoal to meet urban market demands, urban energy policies are working against this option. Urban woodfuel prices in Nairobi are at record highs, but controls on the movement of charcoal restrict transporters from operating outside of a relatively small radius from the city. In areas outside of this radius, woodlots are still commonly established, but they tend to be managed with few labour or capital inputs, and are more seen as a means of maintaining land fallow until such time as other cropping possibilities and sources of capital and labour would encourage households to alter their land-use practices.

8.4 Eucalypt planting and wood markets in India

Introduction

In some areas, tree planting by farmers has been a relatively new innovation, without the long involvement of farmers in wood markets and in markets for land and labour which characterized tree planting in Kenya. The success of some of these recent innovations, judged solely by the extent of new plantings (rather than by their impact on poverty, income, and employment), has been closely linked to the impact of market incentives on household decision-making processes. Eucalypts (*Eucalyptus* spp.) came to be rapidly adopted as a commercial crop in a number of states in India, particularly in the fertile north-west, during the late 1970s and early 1980s. Among the principal incentives for its adoption was the widespread perception that it would be a lucrative crop. This was certainly true for the earliest adopters, but was considerably less so for later adopters who increasingly had to compete in wood markets which were quickly becoming saturated.

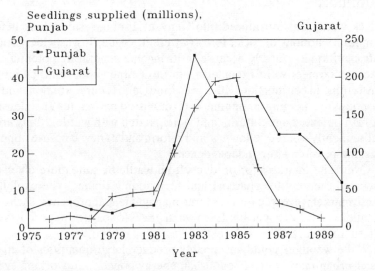

Data for Punjab includes all species.
Data for Gujarat includes only eucalypts.

Fig. 8.5 Uptake of tree seedlings in Punjab and Gujarat states, India, 1975–90.

This section describes the market processes which contributed to the widespread adoption of eucalyptus as a farm tree, as well as the subsequent processes which led to overproduction and to the collapse of the market. Coupled with a number of other features of the rural economy in India, the collapse of the market has meant that growing of eucalypts no longer has the widespread popularity which it once had.

Between 1981 and 1988, farmers in India were reported to have planted around *8.6 billion* trees on private lands. Many of the planting programmes which took off in India were financed by the government, particularly with assistance from the World Bank, and from Canadian and Swedish donor agencies. The speed with which rural households adopted trees, primarily *Eucalyptus* hybrids, into their farming system, was truly phenomenal by any standard. The total newly planted area has been estimated to cover around 2.5 million ha. Most of these new trees were planted either in woodlots, or along field bunds. In western Uttar Pradesh, for instance, around half were planted in woodlots, and the balance were planted in bunds.[6] Seedling uptake in Gujarat and in the Punjab is shown in Fig. 8.5.[7]

The role of market incentives

Indeed, markets for wood products provided a significant driving force which encouraged farmers to plant so many trees. These markets were principally for construction poles, for timber, for pulpwood, and for fuelwood. Suitability for particular markets is partly a function of stem diameter, which is in turn a function of spacing and rotation length. Stems with a diameter of below 10 cm can usually only be used as fuelwood. Wood with a diameter of between 10 and 20 cm can be used as fuelwood, pulpwood, or poles. Stems with a diameter of greater than 20 cm can be used as timber, as well as poles, pulpwood or fuelwood. From the farmer's perspective, eucalypts were well adapted for meeting this range of market demands within the farming system: they grew straight and quickly and with a small crown, which meant that many trees could be planted per unit area, causing little shading on field boundaries.

Certainly among the first generation of donor-assisted social forestry projects in India, the potential impact of markets as incentives for farmer tree growing was almost entirely overlooked by project planners. As with most social forestry interventions which were initiated in the late 1970s and early 1980s, project planning was principally geared to biomass production for woodfuel for subsistence uses. The principal incentives however, were linked to market, rather than to subsistence, demands for woodfuel and other commodities.

Why did markets provide such an incentive for tree planting, and what has been the longer term impact on markets of these interventions?

The first question is quite straightforward. Farmers anticipated huge profits

Fuelwood price (1989 Rs per 100 kg)

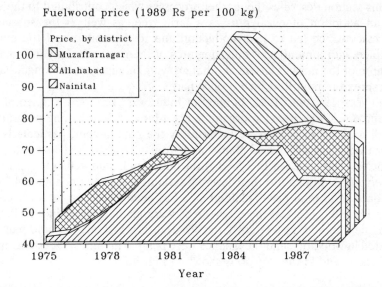

Fig. 8.6 Pricing trends for fuelwood in three districts of Uttar Pradesh, India, 1975–89.

as a result of planting eucalypts. There was a widespread perception of a wood crisis, for fuel and for other types of wood. A study by Bowonder *et al.* (1988) of woodfuel prices in 41 Indian towns showed a 50 per cent increase in prices between 1977 and 1986. Prices for fuelwood in Uttar Pradesh increased significantly, as well (Fig. 8.6). The most lucrative markets, however, were for construction poles. In Uttar Pradesh, construction pole prices more than doubled in real terms between 1983 and 1986. In Haryana, per unit of weight, construction poles were reported to be worth twice to three times as much as fuelwood (Athreya Management Consultants 1989). While woodfuel markets may indeed have been lucrative, construction poles markets were far more so.

For those who managed to get in on the market boom at an early stage, profits were outstanding, and the earliest adopters clearly did quite well. In the mid-1980s, it was reported that a farmer in Gujarat, for instance, was making an annual profit of around Rs 50 000 /ha from eucalypts, making it even more profitable than the best cash crops, principally cotton (Centre for Science and Environment 1985). Internal rates of return were calculated at between 100 and 200 per cent (Food and Agriculture Organization of the United Nations (FAO) 1986).

One of the impacts of this early profitability in growing eucalypts was that it attracted significant levels of investment from particular sectors. Many businessmen and some senior civil servants in the Punjab and Haryana

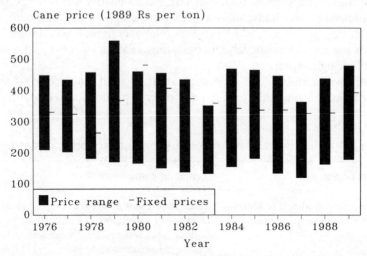

Fig. 8.7 Producer prices for sugar cane in Uttar Pradesh, India, 1976–89. Fixed prices are state-controlled prices paid by sugar factories. The price range reflects the range of prices paid for cane by small-scale jaggery producers.

bought degraded land with a view to using it for eucalypt farming. A number of financial companies bought land with a view to investing in large-scale eucalypt farming using private capital. This transfer of capital from the urban to the rural sector in India was in many respects unique.

Other incentives for tree growing

While potential profits from these lucrative markets did indeed provide an important incentive for farmers to plant trees, the picture was somewhat more complex. Other features of the rural economy gave trees a significant edge over other forms of land use. Prices for other cash crops were particularly volatile, and the perception was that prices for tree products would be far more stable than for the key alternatives, such as sugar cane. Prices for cane fluctuated widely. This was particularly so if producers had to sell cane to small-scale jaggery units for processing, rather than to the larger mills where prices for cane were controlled. Pricing trends for sugar cane in Uttar Pradesh are given in Fig. 8.7, which also shows the range of prices which could be paid by jaggery units. Price fluctuations made it difficult for farmers to evaluate the potential returns from investments in cane production. Trees, they believed, would not be subject to similar fluctuations.

Trees were also viewed as being more resistant to the effects of drought. In Tamil Nadu, for instance, groundnut crops failed every 2 or 3 years. Many poorer farmers converted fields of groundnuts to cashewnut and eucalypt plantations, which could better survive extended dry periods. Conversion also allowed poorer households to seek wage employment elsewhere. In this respect, tree cultivation, which requires relatively fewer labour inputs, can be seen as a means of easing labour constraints. This perspective is discussed in the following chapter.

There was some concern that large areas of good arable land were being planted under trees, at the expense of food crop production. The concern was,

Table 8.5 Objectives and outcomes of social forestry programmes in India

Project objectives	Actual outcome
Produce fuelwood and fodder for local village consumption	Trees were planted for markets, which did little to improve consumption within the village. Fodder trees were generally ignored. Close spacing to accommodate more trees affected grass production. Poles were favoured over other products like fodder and fuelwood.
Reduce pressure on reserved forest land	As projects did little to meet the demand of the poor for fuelwood and fodder, pressure on forest land continued. Moreover, afforestation of reserved forests was a low priority.
Regenerate wastelands, hills, barren slopes and eroded terrains	Trees were planted on farmlands, irrigated land, and land belonging to government. Block plantings on degraded wasteland were given a low priority, and the survival rate was often poor. 'Wasteland Development' was the slogan—tree farming was the reality.
Involve communities in management	Communities were generally not given public funds to raise trees. Instead, Forest Departments raised trees on common lands, which were seen as government trees. The problem of handing over these plantations to local communities remains unsolved.
Help the poor by improving their access to forestry products, and by increasing incomes through asset creation	Participation of small farmers in farm forestry has remained poor. Diversion of land from agriculture to forestry limited employment opportunities. Tribal areas, where people's dependence on trees is more acute were generally ignored. Multi-purpose trees and strengthening the position of 'gatherers' were not objectives. Social security schemes were only implemented on a small scale.

Source: Chambers *et al.* (1989).

that in doing so, food prices would be pushed up, and that households dependent on the market for their food supplies would be further marginalized. Yet for households able to take advantage of markets for wood products, the use of agricultural land for tree growing appeared in some circumstances to make far greater sense than its use for arable crops.

In Karnataka, for instance, even small farmers planted eucalyptus in place of *ragi* (sorghum), production of which was highly variable and unpredictable because of rainfall. In Gujarat, the area converted from food crops to trees was estimated to be no more than around 30 000 ha (out of around 250 000 ha), and this conversion was thought unlikely to exert upward pressures on food prices (Longhurst 1987). Other analysts concluded that up to 50 per cent of the land used for farm forestry displaced other crops (International Labour Organization 1988).

In other areas where wood product markets have been stable, where they appear to provide good opportunities for income generation, and where soil quality was poor, tree planting was a way of making the best of a poor situation. In West Bengal, where the *Patta* Land Scheme was a major instrument for allocating land to the landless, land productivity is typically low, and gross returns from *eucalypts* were far greater than returns to arable crops. Over time, *patta* lands have been increasingly used for tree growing instead of for arable crops (Singh and Bhattacharjee 1991). The use of *patta* land for tree growing is an alternative to taking it out of production altogether, or to renting or leasing it out to another farmer.

A number of writers have critically pointed out that there were clear divergencies between the stated objectives of India's social forestry programmes, and the actual outcome. The huge volume of wood which has been produced, for instance, has had an impact on markets for pulpwood and poles; but neither of these outputs were mentioned as programme objectives. The prevailing view among planners has mostly been that the planting of trees was an end in itself; the outcome of these plantings on household income and employment was never properly evaluated as a means for judging the success or failure of these investments. Some of the gaps between the objectives and outcomes of social forestry projects are summarized in Table 8.5.

The differential impact of markets in encouraging farmers to plant trees

In this regard, a final point relates to the differential impact of markets as an incentive for particular groups of farmers to plant trees. Certainly in the earliest stages of adoption, most trees intended for the market were planted by farmers who were already better off. In Uttar Pradesh, for instance, Gupta (1986) showed that farm forestry had been more popular with relatively large land owners, and absentee land owners, particularly urban businessmen. As tree farming became more widespread, it came to be adopted by a range of farmers. A number of reviews noted the disproportionate involvement of large farmers in farm forestry during the early stages of adoption, but also noted evidence of the increasing participation of small farmers (United States Agency For International Development (USAID) 1988; Arnold *et al.* 1989).

The appeal of social forestry among wealthier farmers was partly because of their better ability to respond to market opportunities. This fact, coupled with the greater ease of managing labour for growing trees, as well as greater risk bearing capacity meant that it was far likelier that wealthier households would adopt extensive tree planting practices than would poorer households.

In Uttar Pradesh, a survey of four villages found that operators of larger holdings were more likely to plant trees (in any configuration, and not just in woodlots) than operators of smaller holdings. Around 52 per cent of farmers owning more than 2.5 ha had planted trees, compared with 17 per cent of farmers owning between 0.5 and 2.5 ha and 5 per cent of farmers owning less than 0.5 ha. Among both large and small farmers, more trees were planted by upper caste households. These findings are of particular interest, especially when considering the fact that in almost all other respects, small farmers followed cropping patterns and used similar levels of inputs, such as chemical fertilizers, as large farmers.

Woodlot planters in Uttar Pradesh also had several times more land, assets and non-agricultural incomes than either other tree planters or non-tree planters. Small farmers who were woodlot planters also had substantial assets, compared with other small farmers. The likelihood that a household would establish a woodlot was greater the more land the household operated, the greater the extent of its assets, and the more diversified were its sources of income. These factors combined to ensure that households would have alternative sources of income to tide them over until trees could be harvested and sold.

Particularly in the fertile north-west (such as in western Uttar Pradesh), the adoption of eucalypts as a land-use innovation closely mirrored the experience of farmers with regard to the adoption of other new crops. Those most likely to adopt these land-use practices were farmers who had the resources to support themselves until their investment paid off. In some respects, increases in agricultural production improved the economic setting for the introduction of eucalypts because of the increased labour and supervision required for more intensive cultivation of annual crops. This is discussed in the following chapter.

These factors *enabled* farmers to adopt woodlots, but the question perhaps remains of what *compelled* them to do so. Markets for tree products in Uttar Pradesh and for factors of production operated in three particular ways which encouraged this transition. First, the profitability of other crops had been declining and rural purchasing power had been eroding since the mid-1970s. The promise of better returns on lower investments thus fuelled the adoption process. Secondly, the great variability of income from other crops meant that they were far more risky to cultivate than were trees, the value of which had been steadily rising over time. Finally, labour costs and farm management problems meant that input costs were much lower for trees than for other crops.

Among small farmers, farm forestry gained some appeal for a number of the same reasons as among larger farmers. In addition though, particularly for households which were able to plant a broader range of tree species in open configurations around homesteads and in fields (instead of in woodlots), small farmers hoped to benefit from increased production of tree products for home consumption such as fodder, fruit and mulch (Arnold *et al*. 1989).

Overproduction and the collapse of the market

The promise of the market failed, in the end, to provide the riches the vast numbers of tree farmers had anticipated. As Fig. 8.6 suggests, prices for fuelwood began falling between 1983 and 1986 and similar falls were experienced in the price of pulpwood and construction poles. Overproduction, combined with a number of other factors,[8] caused the markets to collapse in many areas. The assumption by many tree farmers that there were large markets capable of absorbing future production proved to be incorrect. The range of market options were in fact quite limited, particularly in light of the huge areas planted to eucalypts.

In Gujarat, for instance, between 1978 and 1984 around 600 million seedlings were supplied from nurseries. Even assuming a 60 per cent survival rate, the first harvest of these trees (which were mostly eucalypts) would have produced over 15 million m^3 of roundwood over 6 years, or around 360 million poles (varying from 27 million poles per year in 1986, to 120 million poles in 1991). By comparison, it was estimated that the annual market and subsistence demand for poles in 1988 totalled only 38 million poles. A collapse in the market was inevitable.

It was characteristic of wood markets in India that farmers who speculated by planting trees had no way of anticipating how, where, or at what price their production was to be sold. In contrast, most annual crops in the heavily commercialized north-west are marketed through a sophisticated infrastructure, dependent in part on generous subsidies built into the system of government purchase which absorbs surplus production and stabilizes prices. The lack of such a wood marketing infrastructure, and the problems inherent in designing a system of subsidies and price supports which could even out long-term cycles of wood overproduction and deficit, prevented farmers from being able to make financially sound tree planting investments. As markets came to be saturated, the trend in the late 1980s was for the prices of fuelwood, pulpwood, and construction poles to converge.

It has been argued that as markets came to be saturated, tree growers were subject to exploitation by middle-men and intermediaries, and indeed, in some areas, this seems to be the case (see for instance, Singh and Bhattacharjee 1991). Much farm production is sold to village agents, who sell on to other intermediaries and end-users, and agents play a critical role in the marketing link. The picture is quite complex, however, and the complicity of traders, though not wholly untrue, was not a major cause for the poor prices which farmers were receiving for eucalyptus. Indeed, both large and small farmers surveyed in Uttar Pradesh tended to receive around the same prices. Traders' margins were actually quite small, compared with the other costs of production and marketing.

An unanticipated cost which tree planters have had to bear has been the cost of crop losses. It was originally envisaged that the planting of trees on field bunds would involve low opportunity costs for land, as these areas were otherwise unutilized. Farmers were seldom advised that crops could be affected by competition. None the less, competition for nutrients, light, and moisture often caused yields of adjacent crops to decline. Three years after first establishing trees on field boundaries, many farmers in Uttar Pradesh found they could expect crop losses of between 10 and 25 per cent, depending on soil and water conditions, spacing, location of trees, and other factors.

Farm forestry and contradictions in public policy

Overproduction, marketing constraints, and biophysical limitations all contributed to the growing unpopularity of eucalypts as a farm tree and as a cash crop. Although farmers continued to be provided with often heavily subsidized seedlings from government nurseries, and although farm forestry practices, in response, came to be adopted by wealthy and poor farmers alike, government policies tended to be inconsistent. On the one hand, government sought to increase the incentive for tree planting through seedling subsidies, while on the other, it maintained an interest in keeping wood prices low in the principal and best organized markets: for pulpwood and fuelwood.

Roundwood grown by farmers was suitable for pulping, but paper mills relied more heavily on state forest departments which tended to provide mills with subsidized roundwood. In Uttar Pradesh, for instance, the State Forest Corporation reported that during 1983/84, the cost of raising eucalypts in government plantations was estimated to be around Rs 220 /ton, while it was sold to paper mills for around Rs 140–196 /ton. By comparison, pulpwood which was sold in auction markets was available for between Rs 500 and 700 /ton.

These subsidies have had a major impact on producer prices simply because of the scale of government wood production. Until farmers started harvesting and selling their production in the mid-1980s, government was a monopoly supplier of roundwood. Around 250 000 tons (more than half of the total wood output from Uttar Pradesh forests in 1987/88) was accounted for by state-grown, subsidized, eucalypts. The parastatal Uttar Pradesh Forest Corporation, handles the harvesting and sale of timber. It supplies eucalyptus pulpwood to paper mills at a predetermined rate and timber logs to wholesalers through open auction.

In some areas, rather than moving out of the more lucrative markets for timber and pulpwood and opening them up to farm production, timber production and marketing parastatals are actually *increasing* production for these markets. In Uttar Pradesh, for instance, the Uttar Pradesh Forest Corporation is increasing the proportion of timber it produces *vis-á-vis* less lucrative markets for fuelwood. The supply of fuel-grade wood, as a proportion of total wood produced from forest lands in Uttar Pradesh, has declined from 83 per cent in 1960/61 to 35 per cent in 1989/90. Over the same period, in absolute terms, production of fuelwood has declined from 2.1 million tonnes to 0.26 million tonnes. Even this fuelwood is sold from forest depots at subsidized rates. In the Uttar Pradesh village of Hajeera, for instance, fuelwood could be obtained from forest depots for Rs 150 /ton (about 20 per cent of the market price).

Commercial interests and the monopoly position of the State tend to be protected by legislation. Much forest legislation in India makes the involvement of private producers in the market quite problematic. For example, permits for harvesting and transport are often required, and sometimes, State agencies are the only organizations authorized to purchase forest products. Forest legislation tends to protect entrenched political and commercial interests, and the interests of those who are entrusted with enforcement and who are best able to extract bribes in exchange for the production of the required harvesting and transport permits.

Because of the need for transport permits, tree farmers are less inclined to make arrangements for sale and marketing themselves, and rely heavily on village agents to whom standing trees are sold, and who organize transport. In four Uttar Pradesh villages, around 70 per cent of pre-harvest sales were arranged through village agents principally because farmers were unwilling to assume the risk and cost of arranging transport permits themselves. Farmers and agents both reported that, even if they held valid transport permits from the Forest Department, the police still harassed them and threatened them with arrest on the pretence that the permit had been fraudulently obtained, or that its conditions had not been adhered to. Village agents who dealt with the police on a regular basis were far more familiar with the process of offering the appropriate bribe in order to get their stocks transported to the market. For small-scale tree growers, the costs of arranging the needed permits and of paying the required bribe tend to be prohibitive. This is particularly so when only small volumes are sold. In only seven of 59 cases surveyed in Uttar Pradesh did farmers sell more than 500 trees at a time. Clearly, the system favours agents who are able to combine lots sold by several farmers and thus distribute the costs of obtaining permits.

Impacts of the collapse of the market

All of these factors—overproduction, marketing difficulties, biophysical limitations, and contradictions in government policy with regard to production, marketing, and legislation—have combined to greatly increase the unpopularity of eucalypts as a farm crop. The uptake of seedlings from government nurseries has fallen off dramatically, and in some areas, eucalypt trees are being uprooted. In the Punjab, a senior forest officer summed up the situation,

The prices offered by the traders are no longer remunerative. There is a virtual panic amongst the farmers about the future of eucalyptus plantations raised by them. Some have even started cutting down the young plantations. (Aulakh 1990)

Prices for tree products have fallen significantly in some areas. In Gujarat, the price of eucalypt poles fell by around 36 per cent between 1985 and 1989 (Sharma *et al.* 1991). In one of the largest eucalypt markets in eastern India, in Chandrakona, West Bengal, pole prices fell by around 38 per cent between 1988 and 1990 (Indian Institute of Management 1991). In the Punjab, pole prices in 1988 were reported to be only 15 per cent of their 1984 prices (Khare and Rao 1991).

The uprooting of these vigorously coppicing trees is a very labour intensive task, but some farmers have chosen this option over the cost of facing future crop losses and loss of income because of an inability to sell their production at a reasonable price. In Gujarat, for instance, it was reported in two *talukas* of Mehsana and Bhavnagar districts, that farmers who had felled their trees after the first rotation were planning on returning their land to agricultural uses (Wilson and Trivedi 1988). Among 28 eucalyptus farmers in western Uttar Pradesh, 10 had uprooted their trees while another two had decided to leave only those which had been planted on poor soils. The extent of uprooting nationally has not been properly evaluated, and most reporting of this practice has been anecdotal.

In the long run, the likely outcome of the collapse of markets since the late 1980s and the overproduction which precipitated this collapse will be declining production, the conversion of tree growing areas to other land uses, as well as the development of new markets for tree products. In the long-term, after a series of cycles of overproduction and shortage, some stability in production and in the markets is likely.

Summary

The widespread adoption of farm forestry in India was the clear outcome of its market potential, the anticipation of high future wood prices, and the timely provision of an appropriate farm forestry technology which fit into a particular and well-suited niche in the rural economy. Farm forestry came first to be

Table 8.6 Comparison of wood product markets, and their operation and impact in providing incentives for farmers to cultivate and manage trees

Characteristics of markets and systems of production	Country case			
	Sudan	Kenya	India	
Type of wood and tree product markets	charcoal	wattle bark, charcoal, fuelwood, building poles	building poles, fuelwood, pulpwood	
Sources of supply which meet market demands	agricultural land clearance	farmer-grown *Acacia mearnsii* woodlots	farmer grown eucalytp woodlots	
Competing sources of supply	none	plantation-grown wattle for bark and charcoal, land clearance for charcoal, eucalypt woodlots for building poles and fuelwood	government grown timber supplies	
Time-scale and impact of markets on farmer tree growing	never provided any incentive for tree growing	long-term farmer response allowing scope for the emergence of new markets for tree products	short-term farmer response resulting in overproduction and market collapse	
Other incentives not directly linked to markets	none of relevance; some irrigation is made available for eucalypts and *Acacia nilotica*, but it is not economic to grow these species for charcoal on any scale	seed packets and technical advice provided by Colonial government and tanning extract companies; before Independence, scope for cultivation of other cash crops was limited by legislation	subsidized seedings and technical advice provided	

Table 8.6 (*cont.*)

Characteristics of markets and systems of production	Country case		
	Sudan	Kenya	India
Earliest adopters of tree growing innovations	no adopters	earliest adopters were landed and wealthy, though later adopters included small-scale farmers as well	earliest adopters were landed and wealthy, later spread to other farmers; some planting financed by urban capital
Comparative advantage of sustainable production systems	only comparative advantage would be in terms of proximity to urban markets; high costs of production mean that it is unlikely sustainable systems of production could be competitive	land tenure constraints before Independence encouraged households to plant trees to establish land rights; trees also favoured by households with labour supervision problems or with low demands for income	low costs of labour and supervision *vis-à-vis* annual crops; low profitability of other crops; markets for other crops thought to be more unstable than markets for tree products.

Table 8.6 (*cont.*)

Characteristics of markets and systems of production	Country case		
	Sudan	Kenya	India
Bioeconomic constraints to the adoption of tree growing innovations	low productivity and high costs of growing trees in an arid environment	high opportunity costs for land (tea and coffee more lucrative than trees)	planting on field boundaries caused competition with other crops
Policy constraints to the adoption of tree growing innovations	no clear policy constraints; bioeconomic constraints are more critical	movement controls on charcoal; price controls; paraffin subsidies	movement controls; large-scale and heavily subsidized government production for the most lucrative markets limited farmer access to them
Characteristics of the ways markets operate	single entrepreneurs control most aspects of the market production and delivery system; heavily integrated and sophisticated markets	more segmented market operation, with transporters providing the crucial link between rural producers and urban markets	movement controls favour the emergence of village agents who have the dominant role in linking producers with centres of demand

adopted by wealthier households, but eventually came to be widely adopted across a range of household income classes. Markets provided an incentive for tree growing for three particular reasons:

- First, the profitability of other crops had been declining and rural purchasing power had been eroding since the mid-1970s. The promise of better returns on lower investments thus fuelled the adoption process.
- Secondly, the great variability of income from other crops meant that farmers believed they were far more risky to cultivate than were trees, the value of which had been steadily rising over time.
- Finally, labour costs and farm management problems meant that these input costs were much lower for trees than for other crops.

The immense power of these incentives caused thousands of farmers across the country to become tree farmers. The impact was inevitable: huge quantities of timber flooded the market, depressed prices, and caused an eventual collapse in the market. This, coupled with exploitative practices in the market, as well as unforeseen negative impacts on crop production have prompted farmers to moderate planting practices. Markets are likely to stabilize over the long run, as farmers convert woodlots to other land uses.

8.5 Implications of markets for policies and planning

We suggested at the beginning of this chapter that market interactions ultimately determine how rural households choose to use their land. Markets, in turn, are often the point at which policy makers are able to intervene, and where they can be most effective. The *ways* in which policy makers intervene, however, may be determined by institutions and interests which conflict with what might be considered to be optimal. In the three cases reviewed in this chapter, in Sudan, Kenya, and India, wood product markets, the way they operate, and the policies which have affected them, have had quite different impacts on farmers' incentives to cultivate and manage trees. Some of the characteristics of the three cases are summarized in Table 8.6.

Agrarian policies and political advantage

In understanding the policy dimensions of market interaction, it is necessary to emphasize that the policies which many governments adopt toward tree growing (like the policies which governments adopt toward the agricultural sector as a whole) are often quite fundamentally derived from areas where governments believe they can achieve the greatest political advantage.

Indeed, agricultural policies are often driven by the potential for political gain rather than by the socially or economically optimal. For example, to strengthen the incentive for food production, governments have two main options: either to allow prices to rise in the face of shortages, or to subsidize the costs of farm inputs (including the costs of capital and other factors of production). Both options would have the effect of increasing farmers' profits, thus increasing the incentive for food production. Many analysts have concluded that pricing policies are the more efficient way of securing the objective of greater production; yet governments, particularly in Africa, prefer the latter approach because of its political advantages (Bates 1981).

These types of agricultural policies are geared to rural households. A third option with regard to food markets and prices is directed primarily toward urban dwellers, and the political interests they represent.

. . . Political pressure for low-cost food comes from two main sources. One of course is the urban worker. The other is the employer who, when his workers are faced with high cost food, is forced to pay higher wages. For political reasons, African governments must appease the urban worker; but, as major employers, and as the sponsors of industry, governments share the interests of those who pay the wage bill. To appease consumers while pursuing their own interests, governments therefore join with workers and industry in seeking low-cost food . . . (Bates 1981)

In the face of these pressures, then, rather than allow food prices for urban dwellers to rise, governments choose instead to control prices, and to ration supplies. Rent seeking behaviour on the part of food producers and intermediaries forces the market underground, and parallel markets emerge. Particularly in Africa, over time, this policy has done much to undermine the viability of state marketing structures. Similarly perverse effects can result from the application of similar policies, such as pricing or movement controls, to tree products such as woodfuels or construction poles.

Policies and the implications for rural afforestation

All of these approaches toward the more general area of agrarian policy have relevance to policies which are intended to encourage farmers to plant trees in order to meet the demands of the market. Indeed, agrarian policy has served as the model for social forestry policy in many developing economies. The model though, is again derived more from the potential for political gain, than it is from any clear understanding of how markets operate, and how farmers might respond to a range of market and pricing incentives which would encourage them to plant trees.

For example, if the policy objective is to increase the production of tree-based products to meet market demands, the options are principally to allow prices for tree products to rise to a level where other crops (*vis-à-vis* markets and prices for farm inputs) have less of an advantage over planted trees, or to provide subsidized inputs. Governments consistently choose to provide often heavily subsidized inputs, such as seedlings, rather than to find ways of achieving the socially optimal by allowing markets to define comparative advantage.

Implicit in the assumption that farmers 'need' seedlings is that they know little or nothing about growing trees in the first place. A growing body of research is showing that, in many areas, this is quite clearly not the case. If the objective of subsidizing seedling production is to increase farmers' profit margins (as it is with subsidizing farm inputs to increase the margins to food crop production), then the approach is badly misplaced.

Similarly, where urban markets for woodfuel are hit with 'shortages', energy planners often rely heavily on price controls to keep prices low and on the rationing of urban energy supplies, while at the same time promoting tree planting in rural areas. Price controls and rationing are often accompanied by heavy subsidies to encourage the use of alternative cooking fuels such as paraffin.

Price controls, rationing, and the subsidizing of alternative cooking fuels are practices which closely parallel agricultural policies which place controls on food prices, which ration food, and provide subsidies to keep food prices low in order to satisfy an urban electorate. With regard to woodfuel markets, the outcome may be fuel-switching, and the emergence of parallel markets for woodfuel. As with policies which seek to keep the price of urban food supplies low (policies which have little or no impact on encouraging farmers to grow more food) these types of energy policies do nothing to increase the incentive for farmers to plant trees to produce woodfuels for the urban market.

Policy formulation can be particularly problematic when there are conservation objectives. To rural households, woodlands and forests may appear to provide relatively abundant supplies of woodfuel which can be harvested and sold in nearby towns. This type of harvesting may provide communities with their primary, or only, source of income. While the market plays a crucial role in this respect, the primary incentive which households who are engaged in harvesting woodfuel is for income generation, and not for their own woodfuel consumption. A transition to systems of sustainable production may entail heavy investment costs in several respects: in terms of increasing income and employment possibilities for these households; in terms of developing incentives for local communities to protect valuable ecosystems; and in terms of intervening in the market in a way which encourages consumers who are dependent on woodfuel markets for their supplies to either consume less, to shift to the use of other fuels, or to start using woodfuel which is grown on a sustainable basis (often at considerable cost).

Markets are clearly part of the picture in defining how trees come to be incorporated into farming systems. If we start with the view that trees *do* have a role in these systems and that rural economies are better-off if households can produce sustainable supplies of wood to meet market demands, then the challenge becomes one of how policy and innovation can be introduced and managed in a way which encourages a transition from a reliance on the harvesting and clearfelling of woodlands and forests to meet these demands, to a reliance on economically-viable sustainable systems of household production.

Annex. The financial viability of sustainable charcoal production in Sudan

The ability of multiple suppliers of charcoal in Sudan to compete in the same market is a function of the cost of wood at the source, the cost of burning charcoal and loading it onto transport, and the cost of transport to the market, or:

(1) $P = CW + CC + TR$

where P = the price in the market off-the-truck;
         CW = the cost of wood;
         CC = the cost of burning charcoal and loading it on to transport; and
         TR = the cost of transport to the market.

For a source of charcoal derived from B to compete with a source of charcoal derived from A, it must be true that $P_A \geqslant P_B$.

(2) $P_A = CW_A + CC_A + TR_A$;
(3) $P_B = CW_B + CC_B + TR_B$.

If $CW_A + CC_A$ is less than $CW_B + CC_B$ (for instance if wood at A is acquired at no cost from natural forest clearing and wood from B is grown in plantations, and charcoal production costs are constant), in order for charcoal derived from B to be competitive in the market, TR_A must be greater than TR_B by the difference between the costs of charcoal and wood production to the truck at A and B. In other words, if charcoal can be produced from plantations closer to the market at B, it may have some advantages over charcoal produced at A if it can be produced for savings in transport costs which would otherwise be spent travelling the greater distance to A instead of to B.

Transportation in Sudan is widely available, and because of the distances which are involved, is usually costed on a one-way basis. Transporters seldom travel any distance without a paying load. Transport costs vary with distance. For instance:

(4) $TR = D * T$

where D = the one-way distance; and
         T = transport costs in S£ per ton-km.

It has been shown as well that T varies with distance. Transport costs per ton-km are usually higher for short-distance hauls than they are for longer distance hauls.

A radius of haulage, B, around urban charcoal markets can be determined within which it would be financially viable to produce charcoal which could compete with charcoal produced at *A*. *B* is determined by setting P_A equal to P_B, and then solving for *B*. For instance:

(5) $P_A = P_B$; and

(6) $CW_A + CC_A + TR_A$
        $= CW_B + CC_B + TR_B$;

(7) $CW_A + CC_A + (D_A * T_A)$
        $= CW_B + CC_B + (D_B * T_B)$, or

(8) $(D_B * T_B)$
        $= CW_A + CC_A + (D_A * T_A)$
        $- (CW_B + CC_B)$

But transport costs per ton-km (*T*) are a function of the distance of the haul, as well as of a number of other factors. (*CC* and *CW* may also be partly a function of the distance from urban *labour* markets, but in this analysis, these variables will be calculated independently from distance-dependent transport costs.) A comprehensive study of the cost of transportation in Sudan for major agricultural crops, sesame and sorghum, was carried out in a farm survey in 1984/85 (el Hanan *et al.* 1986). This study pointed out that transportation costs per ton-km in Sudan are primarily a function of:

- regional differences in the supply and demand for transport;
- differences in load size;
- differences in the type of load; and
- the difference in the distance the load is carried.

Regression estimates, based on 347 samples taken during the 1984/85 season yielded the following relationships:

(9) $LN(T) = 1.9341 + 0.1950DDM + 0.3270DHB$
           $- 0.6989LN(D) + 0.0005LN(L)$, for sorghum; and

(10) $LN(T) = 2.0369 + 0.1828DDM + 0.2260DHB$
           $- 0.7072LN(D) + 0.0194LN(L)$, for sesame;

   where $LN(T)$ = the natural logarithm of transport costs
                    in S£ per ton-km;

           DDM = regional dummy values of
                    1 for the Damazine regions and
                    0 for the Gedaref and Dalanj/Habila regions; and

           DHB = regional dummy values of
                    0 for the Damazine and Gedaref regions
                    1 for the Dalanj/Habila regions;

           $LN(D)$ = the natural logarithm of the travel distance in km;

   and

           $LN(L)$ = the natural logarithm of the load size in tons.

For sorghum, the coefficient of determination (r^2) is 0.676 and for sesame is 0.433.

The largest sources of charcoal in Sudan come primarily from the Roseires/Damazine regions. Large-scale charcoal production is shipped to Khartoum and to other markets on 20–ton trucks. For other sources of supply to penetrate the market, they would have to be able to compete with these markets. In fact, transport costs for sorghum transported from this area are very close to reported transport costs for charcoal from the same area (el Faki Ali 1985).

Equation 9 can be algebraically simplified (assuming $DDM=1$ and $DHB=0$) so that

(11) $LN(T) = 2.1291 - 0.6989LN(D) + 0.0005LN(L);$

(12) $T = e^{[2.1291 - 0.6989LN(D) + 0.0005LN(L)]};$

(13) $T = \left[\dfrac{e^{2.1291} * L^{0.0005}}{D^{0.6989}} \right]$

(14) $TR = \left[\dfrac{e^{2.1291} * L^{0.0005}}{D^{0.6989}} \right] * D;$

(15) $TR = e^{2.1291} * D^{0.3011} * L^{0.0005} * Q$

Because these surveys were undertaken in 1984, TR has increased since then because of inflation. Preliminary estimates suggest that a factor of 2.29, roughly equivalent to the high income price index, can be used to update TR to December 1986 prices. In equation 15, Q equals 2.29.

Substituting equation 15 into equation 8 allows us to solve for D_B.

(8) $(D_B * T_B)$
$= CW_A + CC_A + (D_A * T_A)$
$- (CW_B + CC_B)$

(16) $(e^{2.1291} * D_B{}^{0.3011} * L_B{}^{0.0005}) * Q$
$= CW_A + CC_A + (e^{2.1291} * D_A{}^{0.3011} * L_A{}^{0.0005}) * Q$
$- (CW_B + CC_B);$ and

(17)
$$D_B = \left[\frac{CW_A + CC_A + (e^{2.1291} * D_A{}^{0.3011} * L_A{}^{0.0005} * Q) - (CW_B + CC_B)}{e^{2.1291} * L_B{}^{0.0005} * Q} \right]$$

3.321

Assuming lorries typically carry 20–ton loads of charcoal, equation 17 can be further simplified:

(18)   3.321

$$D_B = \left[\frac{(CW_A - CW_B) + (CC_A - CC_B)}{19.2816} + (D_A{}^{0.3011}) \right]$$

If different values are substituted for D_A to reflect the distance between specific urban markets and current sources of charcoal supply, and if D_B is tested for sensitivity to the stumpage rate of sustainably produced wood (CW_B), a range of economic radii of haulage can be estimated.

The manipulation of equation 17, for instance, by including a higher stumpage rate (CW_A) for timber derived from A gives the analyst a view of the potential for making sustainable tree growing for urban charcoal markets more financially viable. Increasing the distance D_A can also give an indication of the *future* financial viability of charcoal production, if estimates can be made either about the distance of future sources of charcoal from urban areas, or about increased transportation costs which would be involved in tapping in to these supplies.

Notes

1. This chapter benefitted from the input of Alex Duncan, Stephen Jones, Mike Arnold, and John English. Much of the material in the second and third sections of this chapter (based on fieldwork carried out by the authors in Murang'a District of Kenya and in Uttar Pradesh State in India) is discussed subsequently from several different perspectives in the chapter which follows.
2. This section is derived from work carried out by one of the authors, reported in Dewees (1986).
3. The balance of production in these areas was burnt by small-scale operators who cleared areas of woodland specifically for charcoal production.
4. The analysis could be quite different if woodfuel were produced as a salvage product from, say, gum arabic plantations, from the management of woodlands, or from irrigated building pole plantations. The production of woodfuel as a single product from woodfuel plantations is likely to be a quite expensive option. It has been assumed that other charcoal production costs remain the same for traditional producers and for sustainable producers. In fact, this is where sustainable producers may be able to gain market advantages: if they are able to produce charcoal more cheaply, if they are able to take advantage of lower labour costs, if they can improve the charcoal market transportation and distribution network. As the market develops in the future, these changes are expected to happen anyway.
5. In the same breath, however, it was noted that the 'inculcation into the natives of an interest in tree planting makes only slow progress'. This is an interesting and fundamental contradiction which is as common today. It is widely acknowledged that farmers have planted many trees on their farms, but it is still claimed that farmers know little about the subject, are unable to plant enough trees to meet their needs, and must be taught to do so and provided seedlings by the extension services.
6. See for instance, Saxena (1991a). Much of this section is derived from fieldwork

by one of authors in Uttar Pradesh State, reported in Saxena (1991a, b, 1992), and has been supplemented with other material.

7. Fig. 8.5 also shows how quickly seedling uptake tailed off as markets for wood products became saturated and as prices, such as those in Uttar Pradesh, shown in Fig. 8.6, also collapsed.

8. These included the exploitation of new sources of fuelwood from large areas of *Prosopis* sp. which had come to maturity.

References

Arnold, J.E.M., Howland, P., Robinson, P.J. and Shepherd, G. (1989). Evaluation of the social forestry project, Karnataka: report to the Overseas Development Administration. Unpublished report. London.

Athreya Management Consultants (1989). Pilot study on experiences in farm forestry in Haryana. Unpublished report. New Delhi.

Aulakh, K.S. (1990). *Economics and economic impact of afforestation programmes in Punjab*. Paper presented at the Centre for Science and Environment workshop on economics of the sustainable use of forest resources at New Delhi, April, 1990.

Bates, R.H. (1981). *Markets and states in tropical Africa*. University of California Press, Berkeley.

Bess, M. (1989). Kenya charcoal survey final report and annexes. Unpublished report. Long Range Planning Unit (Ministry of Planning and National Development), Nairobi.

Bowonder, B., Prasad, S.S.R. and Unni, N.V.M. (1988). Dynamics of fuelwood prices in India: policy implications. *World Development*, **16**, 1213–29.

Centre for Science and Environment (1985). *The state of India's environment, 1984–85: the second citizens' report*. Centre for Science and Environment, New Delhi.

Chambers, R., Saxena, N.C. and Shah, T. (1989). *To the hands of the poor: water and trees*. Oxford and IBH Publishing, New Delhi.

Cowen, M.P. (1978). Capital and household production: the case of wattle in Kenya's Central Province, 1903–1964. Unpublished PhD dissertation. University of Cambridge, Cambridge.

Dewees, P.A. (1986). Urban charcoal markets, pricing, and competition in the Sudan. Unpublished report. Joint World Bank/UNDP Energy Sector Management Assistance Programme, Washington, DC.

Dewees, P.A. (1991). The impact of capital and labour availability on smallholder tree growing in Kenya. Unpublished DPhil thesis. University of Oxford.

Elmer, L.A. (1943). The Kikuyu method of burning charcoal. *East African Agricultural Journal*, **July**, 14–16.

Faki Ali el, G. (1985). *Charcoal marketing and production economics in Blue Nile*. National Council for Research (Energy Research Council), Khartoum.

FAO. (1986). *Tree growing by rural people*, FAO Forestry Paper No. 64. Food and Agriculture Organization of the United Nations, Rome.

Gollop, G.J. (1945). Wattle rules and marketing, 1943–46. Unpublished memorandum from the Assistant Agricultural Officer, Kiambu to the Senior Agricultural Officer (Acting), Kiambu. Kenya National Archives AGR/4/220.

Gupta, T. (1986). Farm forestry. Unpublished report. Indian Institute of Managment, Ahmedabad.

Hanan, M.M. el, Ijaimi, A.L. and Sidhu, S.S. (1986). *Input use and production costs in rainfed mechanized areas of Sudan: results of 1984/85 farm survey*. Ministry of Agriculture and Natural Resources (Department of Agricultural Economics and Statistics), Khartoum.

Indian Institute of Management (1991). Study of major and minor forest products in West Bengal. Unpublished report. Indian Insitute of Management, Calcutta.

International Labour Organization (1988). Employment and income generation through social forestry in India: review of issues and evidence. Unpublished report. ILO Asian Employment Programme, New Delhi.

Kenya, Colony and Protectorate of (1928). *Forest Department annual report, 1928*. Government Printer, Nairobi.

Kenya, Colony and Protectorate of (1929). *Forest Department annual report, 1929*. Government Printer, Nairobi.

Kenya, Colony and Protectorate of (1930). *Forest Department annual report, 1930*. Government Printer, Nairobi.

Kenya, Colony and Protectorate of (1932). *Forest Department annual report, 1932*. Government Printer, Nairobi.

Kenya, Colony and Protectorate of (1935). *Native Affairs Department annual report, 1935*. Government Printer, Nairobi.

Kenya, Colony and Protectorate of (1957). *Price Control Act, Cap. 504.*, Price Control (Woodfuel) Order, Legal Notice 234 of 1957.

Kenya, Republic of (1989). *Development Plan, 1989–1993*. Government Printer, Nairobi.

Khare, A., and Rao, A.V.R. (1991). Products of social forestry—issues, strategies and priorities. *Wastelands News*, **6**, (4), 7–17.

Kinyanjui, M. (1987). Fuelling Nairobi: the importance of small-scale charcoal enterprise. *Unasylva*, **39**, (157/158), 17–28.

Kitching, G. (1980). *Class and economic change in Kenya*. Yale University Press, New Haven.

Longhurst, R. (1987). *Household food security, tree planting and the poor: the case of Gujarat*, Social Forestry Network Paper No. 5d. Overseas Development Institute, London.

Nour, H.O.A. el and Satie, K. (1984). *Charcoal production in Blue Nile Province*. National Council for Research, Energy Research Council (Sudan Renewable Energy Project), Khartoum.

Saxena, N.C. (1991a). Marketing constraints for eucalyptus from farm forestry in India. *Agroforestry Systems*, **13**, 73–85.

Saxena, N.C. (1991b). Crop losses and their economic implications due to growing of eucalyptus on field bunds—a pilot study. *Agroforestry Systems*, **16**, 231–45.

Saxena, N.C. (1992). Adoption and rejection of eucalypts on farms in north-west India. Unpublished DPhil thesis. University of Oxford.

Sharma, K.C., Ballabh, V. and Pandey, A. (1991). *An analysis of farm forestry in Gujarat*. Paper presented at the Institute of Rural Management workshop on socio-economic aspects of tree growing by farmers in South Asia at Anand, India, March, 1991.

Singh, K. and Bhattacharjee, S. (1991). *Economics of eucalyptus plantations on degraded lands: a case study of Nepura village*. Paper presented at the Institute of Rural Management workshop on socio-economic aspects of tree growing by farmers in South Asia at Anand, India, March, 1991.

United Kingdom, Government of (1934). *Kenya Land Commission: evidence and memorandum*, Vol. I. Colonial No. 91.

USAID (1988). Draft national social forestry project mid-term review. Unpublished report. USAID, New Delhi.

Wilson, P. and Trivedi, D. (1987). Eucalyptus: the five year wonder? Unpublished report. Vikram Sarabhai Institute of Appropriate Technology, Ahmedabad.
to 1974

9 Tree planting and household land and labour allocation: case studies from Kenya and India

Peter A. Dewees
N.C. Saxena

9.1 Land, labour and capital allocation

Introduction

We suggested at the beginning of the last chapter that farmers stand at the conjunction of three groups of markets: markets for agricultural commodities, markets for factors of production, and markets for consumer goods. The relationship between supply and demand in these three markets tends to define where the incentive to adopt particular land-use practices might be strongest.

In developing material about the impact of market incentives, the previous chapter showed as well that, with regard to the first group of markets, the decision to plant trees was sometimes related to strong price signals for tree products, such as for building poles and woodfuels. Even when prices are high, however, in any competitive market, multiple producers can only succeed if they can offer a product which has a comparative advantage over other products similarly on the market. In the case of markets for tree products, this comparative advantage is usually defined by the ability of tree growing households to deliver a product for a lower price than the price which other market operators are able to obtain. The cost of bringing that product to the market is going to be defined by the costs of the household's inputs: land, labour, and capital. This chapter explores the relationship between tree growing and this second group of markets: for factors of production.

But a broader perspective is needed. The decision by households to cultivate and to manage trees often has no relationship whatsoever to a market for a particular tree product. Indeed, there may be no such markets at all. Even in the absence of markets for tree products, however, the decision to incorporate trees into farming systems *still* reflects a farmer's perceptions of the costs and benefits of doing so. The values which a farmer places on the necessary inputs and outputs for tree growing may be widely divergent from market values, but the decision to allocate them in particular sorts of ways remains, quite fundamentally, an economic one.

This chapter considers some of the dynamics of factor allocation processes. It begins with a discussion of how labour availability and land-use can be influenced by tree planting (and vice versa), and then develops further material about land and labour interactions in two of the cases discussed in the previous chapter, in India and in Kenya, for which detailed household data on these interactions are available.

Labour-use, land potential, and tree growing

One of the conventional views of the impact of tree growing on employment is that it creates jobs.

. . . Social forestry can give rise to significant employment opportunities for farm families and the landless. These income-earning opportunities are not only in seedling production and in planting, tending, and harvesting trees, but also in complementary activities, such as processing and selling wood and other parts of the tree . . . In situations of high chronic unemployment, this aspect of social forestry can be critical in a strategy for sustainable development. (Gregersen *et al.* 1989)

The converse, which is less commonly acknowledged, is that tree growing can *contribute* to unemployment. Agricultural land which is planted with trees takes far less labour to cultivate and to manage than if it were planted with more labour intensive annual crops. Trees may be appealing to households which are endowed with large land resources *vis-à-vis* other land uses because of the lower labour and supervision costs which may be required to make them productive. In other circumstances, where good wages are to be found in other labour markets, tree planting *can free up* household labour to enable it to engage in better paying jobs elsewhere.

A discussion about the impacts of tree planting on employment (and the impacts of employment on tree planting) must, then, consider how different groups will be affected, depending on their access to sources of labour (both from the household and from labour markets), returns to different land uses at different levels of labour utilization *vis-à-vis* other opportunities for income generation, access to capital to finance the hiring-in of labour, the operation of rural land markets, and the proximity of the household to other labour markets. Households, then, make choices between land uses, reflecting different levels of labour required to make land productive, as well as markets for commodities and inputs which define the margins which households can hope to earn.

Farm studies and the economics of tree growing

One of the difficulties of describing land and labour interactions with regard to tree growing is that most assessments tend to rely heavily on data generated as a result of field station research or on theoretical estimates of input use and of outputs.[1] The reason why such studies tend to rely on theoretical estimates or on field station findings, rather than on data collected from households, is partly related to the problem of measurement. For example, the few days a year which a household might spend on tree planting may be insignificant compared with the tens or hundreds of days spent on other farm activities. The problem of measurement is not limited to the assessment of tree planting inputs. The irregular harvesting of woodfuel and other products from trees is especially difficult to measure in household studies, in contrast with annual crop production, which is harvested and stored or marketed and which can often be estimated with a very high degree of accuracy.

It is often the case that agroforestry technologies are promoted by extension agencies as approaches which can increase farm output, with low levels of inputs. This indeed may be the case, but, coupled with the fact that inputs and outputs are seldom evaluated using real farm data, such assessments tend to view inputs and outputs in isolation from the rest of the farming system. The *unwillingness* of farming households to adopt a range of widely promoted tree planting practices poses a number of problems for the analyst who has suggested that economic returns to these practices pose no serious constraint to their adoption and in fact are so high that they should provide households with significant incentives to incorporate trees into existing land-use practices.

The studies reviewed in this chapter have taken a somewhat different approach by accepting that measurement problems limit the extent to which inputs and outputs can be explicitly evaluated. The view is rather that the pattern of resource allocation among households which have planted trees in particular sorts of ways should be fundamentally different from households which have not adopted these practices, and that these more general patterns of resource allocation can, in the end, be quite enlightening.

The first case examines data gathered in Murang'a District of Kenya. The previous chapter discussed how black wattle woodlots were widely adopted in Kenya in response to market signals, beginning in the late 1920s through the mid-1960s. By any comparison, however, prices for wattle bark which first stimulated widespread adoption, are a fraction of what they once were. Urban prices for charcoal, although at record levels, fail to offer any significant promise of profit because of constraints on harvesting and transporting charcoal, imposed as a result of government's current energy policy. Woodlots, which take up around 6000 ha of good agricultural land in high potential areas of Murang'a, are managed with few inputs of labour or capital and are seldom harvested on any regular basis. The question, really, is why households have chosen to maintain their land under trees, when other crops could yield much higher returns.

The second case considers data collected in a number of villages in Uttar Pradesh in India. As we suggested in the previous chapter, tremendous price incentives for construction poles and woodfuel greatly contributed to the interest of farmers in adopting trees. These incentives, however, had a differential impact, with upper caste households and households which operated larger plots of land being more likely to plant trees than other households. This finding is of special interest, especially when considering the fact that in almost all other respects, small farmers followed cropping patterns and used as much chemical fertilizer as large farmers.

In both Kenya and India, household data strongly indicate that the adoption and maintenance of trees on farms in woodlots is closely linked to a number of issues having to do with labour availability and use, as well as with the extent of household demands for capital. These issues may reflect intergenerational considerations, gender-linked patterns of labour allocation, and quite subtle distinctions in the quality of the land which is used for tree growing.

9.2 Woodlots, labour and land in Kenya*

Introduction

We pointed out in the previous chapter that large areas of land in high potential areas of Kenya are maintained under planted tree cover. Woodlots account for a substantial proportion of the total area planted with trees. In Murang'a District in Central Province, these are principally of black wattle, although in other areas, particularly in western Kenya, species such as *Eucalyptus* spp. are predominant.

The view that constrained labour supplies may have encouraged tree planting in Kenya poses a number of interesting contradictions. First, Kenya has experienced a particularly high rate of population growth over the last two decades, and to suggest that labour could be constrained in these circumstances seems paradoxical.

Secondly, in light of these heavy population pressures, the view that land should be used for growing trees instead of food crops or more lucrative cash crops challenges the accepted wisdom. It is often argued that, in the face of growing demands for land in areas of heavy population pressures, forests are cut down, trees are depleted, soils are destroyed, and the unavoidable cycle of environmental destruction, culminating in widespread famine, is precipitated. While large areas of forests have been cleared to make way for agricultural expansion in Kenya, farmers have subsequently been quite successful at cultivating and managing trees around their farms and in their fields, thus contradicting the conventional view that they are environmentally irresponsible. Indeed, the particular care with which farmers plant and manage trees around their farms suggests that they are quite familiar with the management of their local environments.

Both of these seeming contradictions, that labour is somehow constrained

* This section is taken from Dewees (1991).

in an economy with a large population and with a high population growth rate, and that tree planting is common even in areas of heavy pressure for land, need to be addressed in turn.

A number of studies have suggested that some of the widely held perceptions of the smallholder economy have been entirely wrong.* It is generally argued that there are serious problems of unemployment in Kenya, and that, with a population growing at a rate of nearly 4 per cent in some areas, these problems will become much worse. There is a growing body of empirical evidence which suggests, however, that rural labour supplies have been greatly constrained and have indeed been *constraining* the ability of smallholders to use their land more productively. These constraints are partly linked to serious problems in the operation of land, labour and capital markets.

Particularly among resource-poor families, some households have the special problem of capital availability. While many types of land use can be lucrative, some farmers are excluded from these opportunities because they are unable to raise the required capital to invest and are unable to cover the high recurrent costs which are necessary to get the highest returns on their investment. Similarly, among ageing households with tighter labour supplies, and with fewer *demands* for capital, tree growing may be seen as a means of productively utilizing land which is not really needed to maintain the household until parcels are further subdivided and passed on to the next generation.

One outcome of rural labour and capital constraints in Kenya is the predominance of labour migration. Migration in search of wage employment in urban areas or in plantation agriculture is an attractive alternative for households with limited access to capital or to farm labour. This further complicates the labour supply picture, but provides important sources of off-farm capital for on-farm investment.

Tree growing may be an attractive land use to those households which have problems of either capital or labour availability. Trees require low levels of capital to establish and maintain and can produce income for households which might otherwise be excluded from growing cash crops because of a lack of access to investment capital. Because labour inputs for trees are also low, they may be maintained or adopted by households which find they lack the labour power to cultivate more intensive crops or which have less of a need to maximize the profitability of land use.

Household studies were undertaken in Murang'a District in Kenya to explore how households which operate holdings with woodlots on them are different in terms of their land and labour allocation strategies than other households. A subsample of woodlot growing households was selected from recent low-level aerial photography. A second subsample was developed by identifying, from aerial photos taken in 1967, holdings which were used for cultivating a woodlot in 1967, but which have since been cleared and are

---

* See for example, the discussion in Collier and Lal (1986).

being used for something else. The objective of comparing these two groups of households was to track entry and exit into woodlot establishment and management, as well as key differences between these groups of households.

A survey was carried out of these two groups of households in early 1990. The survey compiled information about factor endowments and resource allocation processes, particularly:

- *household composition*: size, the extent of the resident household, non-resident relatives of resident household members; age, education; children in school; dependency;
- *labour composition*: gender; labour hired in (permanent, casual, seasonal); labour hired out (urban formal or informal, rural agricultural or non-agricultural); seasonality; remittances to the household;
- *general features of land use*: period of residence, additions to the shamba, additions of other landholdings, the extent of local land purchase/rental; cropping patterns; changes in cropping patterns; crop marketing;
- *livestock*: number; meat or dairy; grade, cross-bred, traditional;
- *extent and quality of farm assets*: building quality, number; machinery, equipment, bicycles, etc.; recent changes in assets, how financed;
- *household savings*: use of banking or non-bank financial institutions; role of Savings and Credit Cooperative Organizations; involvement in cooperative marketing, input, and processing organizations;
- *household investments*: sources of finance for investments in the farm (borrowing, land sales, remittances, sale of assets).
- *harvesting and processing of woodlot products*: charcoal, fuelwood, poles, wattle bark; the extent to which these were sold or used by the household; seasonality of harvest; who harvests; disposition of income.

Parcels were measured and land uses were recorded at the time of the survey. The altitude of the holding, and the slope of each plot was also recorded.

Household composition, age structure and employment

The household survey found significant differences in the numbers of people resident in the sampled households. Households which operate parcels with woodlots on them support 5.6 residents per holding. Households which operate parcels which are no longer used for growing woodlots support 6.5 residents per holding ($t=-1.79$, significant to 10 per cent).

These differences are mostly accounted for in terms of the numbers of resident and non-resident children in the respective households. There were no significant differences in the *total* numbers children in sampled households, but rather in the proportion of these who were residents and non-residents. Former woodlot growing households, on average, have one more resident child per household, while woodlot growing households have, on average, one more non-resident child per household.

Differences in the numbers of resident household members are partly related to differences in the age structure of the respective households. Woodlot-operating households tend to have an older age structure. Male heads of these households are around 7 years older than male heads of former woodlot growing households. Female heads are around 10 years older, and wives are around 5 years older. Because these household heads are older, it could be expected that the ages of their children would also be older, and indeed this is the case.

As children grow older, among Kikuyu households, they tend to move away. Because households in the respective subsamples have different age structures, it could be expected that differences would be evident in the numbers and ages of children who are no longer in residence: older, woodlot growing, households could be expected to have a larger number of their children living away from the shamba, and younger, former woodlot growing, households could be expected to have a larger number of children living at home under the age of marriage.

These ideas are borne out by data about composition and age of non-resident relatives of resident household members. On average, there are around 2.30 non-resident relatives (mostly sons and daughters) for every woodlot growing household, and only 1.20 non-resident relatives for households which formerly grew a woodlot, the difference being statistically very significant ($t=3.04$, significant to 1 per cent).

Patterns of land use and distribution

Generally, woodlot growing households operate larger holdings than households which are operating parcels which have been cleared of their woodlots. The mean reported holding size (comprised of all operated parcels) among woodlot growing households was 2.2 ha ($\sigma=1.6$) compared with 1.6 ha ($\sigma=1.6$) among households which operate parcels which had been cleared of their woodlots since 1967 ($t=1.89$, significant to 10 per cent).[2]

A total of 74 different land uses were recorded during the survey, reflecting the large number of different types of intercropping strategies undertaken by farmers. These categories were subsequently grouped into eight more general land-use categories. The objective was to cluster land uses by characteristics which reflected the local view of crops as labour intensive seasonal ones (in Kikuyu, *irio cia īmera*, 'sprouting foods') like maize, beans and vegetables or as the more labour extensive perennial crops (*irio cia menja*, 'digging foods') such as yams, arrowroot, cassava and bananas.

Though woodlot growing parcels are larger, and despite *proportionate* differences in land use, the actual areas of individual land uses (with small exceptions) are statistically similar between groups of households. The only significant differences in land use between sampled parcels are in the area of woodlots and the area under pasture. Woodlot operating households keep over twice as much land (proportionately 60 per cent more) under pasture and fodder crops which on average translates into another 500 m² of pasture or fodder crops. They also manage, on average, another 0.31 ha of woodlots than households which formerly operated woodlots. This basically means that the income generating potential, at least in terms of land use, of the remaining food and cash crops are roughly similar between groups of households.

Having said this, however, tea and coffee, the principal cash crops in this agroecological zone, are managed more intensively and more productively by former woodlot growing households. In comparing net income from tea and coffee, woodlot growing households received a net income from tea which was around three times their net income from coffee. Former woodlot growing households received a net income from tea which was 6.8 times as much as their net income from coffee. If households operated parcels which were similar in site quality and in the way in which labour and capital resources were allocated, it would be expected that these ratios would be roughly similar. The large difference between these ratios suggests that former woodlot growing households operate parcels which are of a fundamentally better quality, and in a way which has a better income generating potential than households which operate parcels with woodlots on them.

The intensity with which particular crops are managed is partly a function of labour availability. If labour availability, defined in part by the structure of the resident labour supply, is compared with patterns of land use, there is less household labour available among woodlot operating households. In short, although households have similar patterns of land use (with the exception of the area under wattle and pasture), they have less resident labour to cultivate annual and perennial crops with any intensity. Nor does the data show that these households tend to rely on hired labour to fill the gap.

A number of other indicators suggest that woodlot growing households are more inclined to use their holding more extensively than others. These households have invested heavily in cattle, the value of their cattle being 40 per cent higher than amongst former woodlot growing households. A larger area of their holdings is under managed pasture. Their use of crop inputs is generally lower, and is more likely to be financed with regular crop income, rather than by credit mechanisms which are available through local cooperatives. Finally, non-resident relatives of woodlot growing households remitted income to the household with greater frequency than other households, suggesting both a greater dependence on remittances among these households, as well as greater availability of these remittances.

Most households operated more than one parcel. There were no differences between groups of households with regard to the number of parcels which were operated. There were also no significant differences in sources of rights of tenure to parcels operated by sampled households. The bulk of the parcels acquired by both groups of households were obtained through inheritance, rather than through purchase or rental. Despite what appear to be age-related differences in household composition, there is no evidence that the acquisition of larger holdings among woodlot growing households is a function of age, that is, older households are no more likely to have purchased additional land than younger households. Woodlot growing households simply inherited larger parcels of land, and this is an important difference between groups of households.

Variability in agricultural conditions

Two other variables which were included in the data set were the altitude of the sample parcel and slope. Differences in both of these variables between subsampled households were found to be significant.

Differences in altitude Subsamples were selected in a way which was intended to limit agroecological variability among households by using the 1800 m contour to help delineate the study area. Among sampled parcels, the mean altitude was 1838 m (σ=51.08). Despite the effort to limit the sample to parcels with similar agricultural conditions, altitude between sampled households varied significantly. Parcels which were being used for growing woodlots were generally lower in elevation (1829 m) than parcels which had been cleared of their woodlots (1848 m). The difference between the mean altitudes is small (less than 20 m) but is significant to less than 5 per cent (t=−2.18).

This is an interesting point. On the one hand, wattle does much better at higher altitudes and so one would expect to find parcels with woodlots to be found at a higher altitude than other parcels. On the other hand, certain other land uses are also better-suited to higher altitudes. Tea, for instance, grows quite well in the higher zones in Murang'a. The fact that cleared parcels are at a higher elevation may be a reflection of this fact: parcels which are lower in elevation are used for growing woodlots while parcels at higher altitudes are more likely to be cleared of their woodlots and used for growing something else.

Differences in slope Slope evidently plays an important role in farmers' decisions to use certain plots for particular land uses. Among all sampled parcels, food and fodder crops and pastures are generally found on the least sloping hillsides, with household compounds found on fairly flat land. Coffee and tea are grown on steeper slopes, and the steepest slopes are used for growing wattle.

Differences *between* sampled households were also tested. For most land uses (annual, perennial and permanent crops as well as household compounds) the slope of plots on woodlot growing parcels are consistently greater than on parcels which were formerly used for growing woodlots (though significantly so only for tea and for the household compound). Slopes were also weighted by the land area used for particular crops, and aggregate slopes were calculated for sampled households. If these aggregated slopes are considered, parcels which are used for growing woodlots are steeper (by around 4°) than parcels which were used for growing woodlots but which have been cleared.

Despite the fact that woodlot growing households operated larger parcels, the quality of this land was generally poorer than among other households. *Independent* of the labour endowment, fundamental differences in the quality of the basic land endowment and in how this is different across households (principally that parcels operated by woodlot growing households are more steeply sloping and are at lower altitudes) have important implications with regard to the intensity with which holdings can be cultivated in the first place.

It could be expected that woodlots would be maintained on steeper slopes to limit soil erosion problems and because it is more difficult to cultivate other crops on steep slopes. These households, however, may have few other options: poorer overall site quality may be acting as a constraint, along with access to other factors of production, in preventing some households from using their land for more highly valued crops.

The differential use of steeply sloping land for tree growing may in part be an outcome of colonial land-use legislation and its post-Colonial successors such as the Agricultural Act and the Agricultural (Basic Land Usage) Rules which regulate hillside cultivation. The Land Usage Rules prohibit the cutting of trees and the grazing of livestock on hills with slopes of greater than 19°. Among surveyed households, the average slope of plots used for growing woodlots was nearly 22°. For most land uses, the slopes of plots for all crops grown on parcels which were also used for growing woodlots are consistently greater than on parcels which were formerly used for growing woodlots, indicating a very basic difference in site quality.

One outcome of the difference in site quality is that woodlot growing households would require relatively larger labour inputs to make their parcels more productive. For example, the labour required to harvest tea would be quite different among surveyed households. Suppose farmers in the respective subgroups each cultivated parcels of tea which were 30 m by 100 m in area (0.3 ha, as in the survey). Suppose also that these parcels were on hillsides with slopes of 18.9° for woodlot growing households and 15.4° for former woodlot growing households (also as in the survey). The difference in total vertical altitude between these parcels is around 6 m. The 2,500 tea bushes which would be planted on such a parcel, would yield around 3,100 kg of green leaf per year, and would require the removal of 210 basket loads of green leaf per year, probably to the top of the parcel. Clearly, the additional vertical altitude of 6 metres for woodlot growing households would result in a very substantial increase in the amount of work required to harvest tea from the parcel.

Discussion

The earliest adopters of wattle found it attractive *because* they lived and worked away from their land. Others found it appealing because it *allowed* them to live and work away from their shamba and to develop businesses elsewhere, while at the same time, generating an income. As Cowen (1978) has suggested, 'During those lengthy periods when landowners . . . . were engaged in wage employment outside of the reserve, the irregular seasonal application of labour permitted the production of wattle without the continuous presence of labour power'.

Indeed, the labour-extensive nature of black wattle largely accounted for its rapid adoption. A Senior Agricultural Officer in Central Province noted, in the early 1940s, people had found out that wattle trees

. . . grow quickly while they are sleeping, so that it is an easy way to get money. They plant the wattle, tend it for a year or so, then go away to Nairobi and find work for some years and know when they return that their 'bank' has grown to be worth many shillings without the owner doing much work.[3]

Early modelling of optimal resource allocation strategies in Central Province clearly indicated that labour constraints could have a significant impact on land use (Clayton 1961).

The picture which emerges from the data about the contemporary situation is quite a sophisticated one which is generally consistent with the view that woodlot growing households are more inclined to manage their holding extensively, partly because of both labour and capital constraints, but for a number of other reasons as well. Woodlot growing households have an older age structure, and consequently have fewer children living at home. The fact that there are fewer residents of these households, and that holdings of woodlot growing households are larger has meant that labour-to-land ratios are lower. It has also meant that their demands for capital are likely to be lower than among other households.

These households have invested more heavily in livestock, and spend relatively little on crop inputs. They depend primarily on regular farm income to cover investment and recurrent costs and are less dependent on cooperatives to provide credit for farm inputs. Urban wage remittances probably help to reduce these households' dependence on cooperatives for loans as non-resident relatives of these households remit income with greater frequency than non-resident relatives of former woodlot growing households.

Some writers have argued that the maintenance of farm trees, in configurations such as woodlots, is a form of risk management.* Unlike virtually any other crop, trees can be harvested whenever the household's needs for cash are the greatest. They reduce the household's exposure to risk because

* See for instance Chambers (1988) and Leach and Mearns (1988).

capital and labour outlays for tree growing are quite low. If one accepts this argument, woodlot establishment would be consistent with other extensive patterns of land, labour and capital use which seem evident among risk averse households.

As we have pointed out, these types of resource allocation practices are also common among older households with fewer demands for capital. Investments by these households in livestock can be easily recouped if capital is needed. By adopting less intensive forms of land use, while still providing an income which is adequate for supporting the household, woodlot growing households are less exposed to the problems of obtaining the capital and labour needed for a high input/high output approach. These households also have greater labour constraints, because children have left the shamba, and so the establishment and maintenance of woodlots is seen by these households to be a low input and labour optimizing land-use strategy. Woodlots are maintained not so much as a source of income (although income generation from woodlots remains a clear possibility), but because household resources appear not to be adequate, or are otherwise not needed, for cultivating these areas more intensively.

9.3 Labour issues and resource ownership among tree farmers in north-west India

Introduction

The previous chapter argued that the rapid and widespread adoption of woodlots in India was driven by very strong market signals, and high prices for construction poles and woodfuel. These incentives, however, operated in conjunction with structural characteristics of agricultural labour markets in rural India, which made the use of agricultural land for tree growing an attractive option, particularly among those households which were constrained from supervising and/or hiring labour-in to more intensively cultivate holdings.

Large-scale tree planting in the north-west of India was neither a response to fuelwood scarcity nor was it linked to particular traditions of tree farming. Shifts in rural land use have to be seen in the context of agrarian change brought about by the success of the 'green revolution' in that region which compelled landowners to adopt resource allocation strategies which reduced labour supervision and farm management costs. Large farmers had a greater compulsion to change their land use. Their comfortable land and asset position and their upper caste status enabled them to risk the adoption of a new land use such as a long-gestation tree crop on their farms.

Labour and supervision issues   Of all the agrarian regions of India, agricultural production rose the fastest in the Punjab, Haryana, and in western Uttar Pradesh. Increased agricultural production, and the intensification which brought it about, introduced a range of problems with regard to labour supply. Intensification was brought about by the widespread introduction of irrigation, coupled with the introduction of improved and higher yielding 'modern varieties' of grain crops and of high value crops such as potatoes and sugar cane. These improvements greatly increased demands for agricultural labour because of the greater number of cultural operations, and because of the simple fact that yields could only be doubled or tripled if there was sufficient labour with which crops could be harvested.

The green revolution in the north-west brought with it a rapid growth in non-agricultural employment opportunities. This introduced new problems of labour supply, in a region far less densely populated than other regions of India. In farming areas where demands for agricultural labour were growing the fastest, villages experienced serious out-migration in response to opportunities elsewhere. Many farmers, too, developed urban interests and moved away from the villages, while still retaining their agricultural lands. In order to cope with serious local labour shortages, and the lack of labour available on short notice, farmers would bring in labourers from the eastern regions. The north-west consistently maintained higher than average levels of agricultural wages (Lipton 1985; Basant 1987; Jose 1988).

Landowners have responded to the changing labour situation in a number of ways. Mechanization, of course, plays a key role (Byres 1981), and this is often coupled with new contractual and tenancy arrangements. Employers often provide credit to labourers during the low season in order to ensure their availability during the peak season and to even out problems of labour supply (Bardhan and Rudra 1978). Landowners may negotiate with labour contractors who will provide labour for operations like sowing and harvesting (Patnaik 1986).

New forms of sharecropping arrangements, similar to traditional tenancy arrangements, are emerging for the more labour-intensive crops. Tenants participating in these arrangements are usually responsible for all manual tasks, while the landowner provides the needed inputs and means of production (Rutten 1986). Tenants are virtually piece-rated labourers, but their involvement helps the landowner save on supervision.

A number of operations are difficult to delegate to hired labourers, such as negotiating with the bureaucracy for electric power or canal water, or running a tubewell or tractor. When these tasks are delegated, landowners can expect to sacrifice some efficiency.

Outright leasing arrangements are sometimes resorted to, though these are not legal. They also entail the risk of loss of land, since such arrangements may confer occupancy rights on the tenants. Prosperity in the north-west has made tenants aware of their rights, particularly increasing the risk for non-resident landowners, who may be unable to screen prospective tenants.

The use of hired labour tends to increase labour management problems as the area under cultivation is increased. Particularly where there are family labour constraints, which limit the ability of the landowner to supervise hired labour, crop cultivation can be especially expensive. Eucalypts were introduced within this setting in the rural economy. In comparison with other crops, labour requirements for tree growing are quite low, and returns were expected to be especially high.

Certainly in the earliest stages after its introduction, it was viewed to be an especially risky crop and its adoption was partly dependent on the willingness of farmers to risk planting it. Risk taking behaviour varies between regions and classes, and has sometimes been related to assets or income, and extent of market participation. It could be expected that the area planted with eucalypts would be higher in commercialized regions, and on larger holdings. It was precisely these holdings which were encountering the most serious labour supervision constraints.

Farm size and issues of risk The planting of eucalyptus on fertile lands either as a woodlot, on field boundaries, or intercropped with seasonal crops, entails some opportunity cost. It takes around 6 years before the first harvest could be expected to generate any income, and households with serious income constraints would be unlikely to see any other revenues in the meantime. Because of this, in some circumstances, tree planting investments are more likely to be undertaken by large farmers with more than just-adequate farm resources, as they can wait longer for the benefits of tree growing to accrue than can owners of small farms.[4]

Moreover, depending in part on access to capital and labour, and on the viability of alternative land uses, tree planting can, in some circumstances, involve a great deal of risk for the farmer. Seedlings may not be available at the right time from the Forest Department. Seedlings may not be healthy enough to grow into robust trees. Eucalyptus grown as a cash crop needs to be marketed. To do so, the farmer has to locate the market, gain access to it, and obtain permission from the Forest Department and from other government offices to transport and sell his trees. Alternatively, he must sell his standing crop to a contractor, who negotiates with the bureaucracy.

In addition to farm size, education and access to information may influence farmers' attitude to risk. In rural India these attributes are often correlated with caste (Beteille 1969; Harriss 1972). Moreover, ownership of productive assets such as milch-cattle or tractors, and diversified sources of income, may give a farmer greater control over the production and marketing environment, enabling him to risk a new crop like eucalyptus.

Given various risks and costs, it could be expected that labour-constrained farmers in general, and non-resident farmers in particular, as well as upper caste farmers with a comfortable resource position would find the growing of eucalypts attractive in commercialized regions where the opportunity cost of land, and the consequent risks associated with planting trees, was highest.

These hypotheses were tested by collecting data from six villages in Uttar Pradesh. Four villages were identified in the districts of Muzaffarnagar and

Table 9.1 Numbers of trees planted, compared with numbers of tree planters

Type of planting, by residence of planter	Number of planters	Number of trees planted (thousands)
Woodlot plantings		
Resident planters	36	29
Non-resident planters	29	136
Boundary plantings	180	107

Nainital, both in the 'green revolution' belt of western Uttar Pradesh, where commercial growing of eucalypts had been practised for about 10 years, over a full production and marketing cycle. The main crops were sugarcane, wheat, rice, potato, and sorghum (for fodder). Two other 'control' villages were selected from the Allahabad district in eastern Uttar Pradesh, which is more subsistence oriented. The main crops in these villages were rice, wheat, pulses, and coarse grains. The results of the survey are discussed below.

Labour and supervision amongst tree-growing households

Data on the use of agricultural labour for the cultivation of main crops and for eucalypts were quite revealing. Eucalypt cultivation and management was found to require relatively little labour, compared with other crops, using only 50–70 mandays of labour compared with 250–300 mandays for annual crops per ha per year. Most of this labour (between 80 and 90 per cent) was required either in the first year or in the final, harvest year. Farmers reported that the need for the intensive supervision of labour during the tree harvests was greatly reduced because production was far easier to quantify and estimate than crop output, and was, consequently, less likely to be stolen by labourers. An important labour constraint for potential tree farmers was that the planting season for both trees and *kharif* crops was common.

Tree planting by non-resident farmers   In the more subsistenc oriented Allahabad villages, only 16 per cent of non-resident farmers took to tree planting, compared with 76 per cent of non-resident farmers in the more commercialized villages of western Uttar Pradesh. In these villages, the number of trees planted by non-residents was almost 120 times the number planted in the Allahabad villages. Table 9.1, based on a census of trees in the four western Uttar Pradesh villages, gives both the numbers of planters, as well as the numbers of trees planted, for the three principal categories of planters: non-resident woodlot planters, resident woodlot planters, and those who planted on field boundaries and around homesteads. Boundary planting proved to be the most common tree planting practice among all farmers, but, in terms of the total *numbers* of trees, most trees had been planted in woodlots by non-resident planters.

Tree planting by resident farmers The cultivation of annual crops in western Uttar Pradesh is done under conditions of secure irrigation and well developed markets, and enjoys a high degree of stability of both production and prices. The use of fertile land under these conditions for the cultivation of woodlots poses some interesting questions.

What distinguishes woodlot planters from others? Part of the answer may lie in problems relating to the deployment of family labour. We argue that the ability of households to supervise hired labour or to provide this labour themselves is a key constraint which significantly affects the households' decision to establish a woodlot or to continue cultivating annual crops.

In order to test this view, we calculated an index which was intended to reflect the extent to which households are able to supervise farm labour or to provide farm labour itself. Most supervisory and labour tasks are the responsibility of male household members. A supervision index was defined to be equivalent to the ratio of land-owned to the number of male working-age household members. 'Supervision-constrained households' were those with a high supervision index.

The value of the supervision index for different planting groups and farm size-classes is given in Table 9.2. Woodlot planters were consistently more

Table 9.2 Supervision indices for different categories of resident farmers

Type of planter	Number of farmers		Mean value of Supervision Index, by farm size class		
	Large	Small	Large	Small	Total
Woodlot planters	12	3	4.48	1.18	3.82
Other planters	56	33	2.94	1.15	2.28
Non-planters	22	9	2.15	0.89	1.78
Total	90	45	2.95	1.10	2.34
One way analysis					
F-value			7.74	0.66	7.93
F-probability			0.00	0.52	0.00

Table 9.3 Supervision indices compared with wage payments and numbers of trees planted

| Supervision Index | Number of cases | Annual payments, in Rs/ha for | | Trees planted per household |
		Permanent labour	Casual labour	
< 1.0	31	188	1153	290
≥ 1.0 but < 2.0	47	634	1787	457
≥ 2.0 but < 4.0	44	510	1734	558
≥ 4.0	15	1011	2385	877
Total/average	137	535	1692	498
One-way analysis of Supervision Index				
F-value		3.01	8.04	5.34
F-probability		0.03	0.00	0.00

supervision-constrained compared with other planters and non-planters, and were statistically so when all size-classes were considered together, as well as for large farmers. Supervision indices among small farmers, were, however, not statistically significant, regardless of whether they were planters or not.

Supervision indices are compared in Table 9.3 with the numbers of trees planted, and with the amount households spend on hiring permanent and casual labour to test the view that supervision-constrained households both hired more labour and planted more trees. The findings in Table 9.3 confirm this view.

Woodlot planters continue to cultivate labour intensive income generating crops, like sugarcane, paddy and potato, as well as their trees. What is characteristic, however, is that new forms of sharecropping arrangements are emerging for the more labour intensive crops. As Srivastava (1989) has suggested, 'From leasing as a means of rental appropriation, landlords have moved to leasing as a means of control over labour power, the appropriation of its produce, and as a means of further accumulation.' However, the technical requirements of a crop were influencing contractual arrangements. Tenants who sharecropped paddy and potato received less than one-quarter of the crop, while labour extensive crops like eucalypts were more commonly grown using hired labour under personal supervision.

Land, resources and risk

Although cropping patterns within the surveyed areas differed from village to village, within a village both small and large farmers preferred similar crops, and put similar proportions of their areas to the main crops. The differences between large and small farmers in both input-use as well as in total production were not significant in many villages. Differences in the use and payment of hired labour, as well as in the proportion of total production marketed were more significant. There were also significant differences between planters and non-planters with regard to the amount of land owned, caste, assets, and non-agricultural incomes.

Tree planting and holding size   The amount of land owned and numbers of trees planted by different categories of households are shown in Table 9.4. On average, the 1617 resident households in six villages owned 1.59 ha per rural family. Sixty-three per cent of the total arable land was owned by large farmers (those owning more than 2.5 ha), and these accounted for 19 per cent of the population. Combined with small farmers (those owning between 0.5 and 2.5 ha), the two categories of farmers owned 97 per cent of the cultivated land, and accounted for 59 per cent of the village population. The other 41 per cent of the population consisted of the completely or nearly landless, owning less than 0.5 ha per household, and accounting in total for only 3 per cent of all

Table 9.4 Tree planting and land ownership

	Farm size class				
	Large (> 2.5 ha)	Small (between 0.5 and 2.5 ha)	Near landless (between 0 and 0.5 ha)	Landless	Total
Total number of households	300	651	248	418	1617
Number of planters (per cent in size class)	156 (52%)	112 (17%)	13 (5%)	9 (2%)	290
Number of non-planters (per cent in size class)	144 (48%)	539 (83%)	235 (95%)	409 (98%)	1327
Number of trees planted, per household	394	53	13	3	97
Area of land owned (ha)	5.4	1.4	0.3	nil	1.6
by planters	6.0	1.6	0.4	nil	3.9
by non-planters	4.7	1.3	0.3	nil	1.1
t-values, land ownership, planters and non-planters	3.1	4.8	2.3		18.7
significance level	0.002	0.000	0.024		0.000

cultivated land. Many of the households in these categories were agricultural labourers, or depended on other wage work for their subsistence.

Out of 1617 resident households, 290 had planted 50 or more trees since 1980, and were classified as planters. Their distribution among the four types of households was heavily skewed: more than half of all large farmers were planters, while less than 4 per cent of landless and nearly landless farmers were planters. Small farmers, in contrast, accounted for 40 per cent of the population and owned 34 per cent of the land, but had planted only 21 per cent of all trees.

Table 9.4 shows that in each category of land ownership, planters owned more land than non-planters, and that these differences were highly significant. If the category of planters is further divided into woodlot planters and others planters (who planted on bunds and around homesteads without displacing crops), the average number of trees planted by these two categories of households was 1250 and 367 per household, respectively. Tree planting and land ownership are significantly correlated. On average, woodlot planters, other planters, and non-planters owned 6.35, 3.41, and 1.09 ha respectively (F value = 195).

Even within farm size class, tree ownership proved to be highly skewed. Among small farmers, for instance, the top 10 per cent of small farmers owned 75 per cent of the trees within this class. Farm size class distinctions fail to account for joint family holdings where land has been partitioned between sons in order to evade land ceiling laws. Such families often owned more than one house in the village, and, although different members of the family ate separately, they were in charge of the same production unit, or shared capital assets, such as a tractor.

In addition to categorizing farmers on the basis of size-holding, we considered other elements which account for differentiation, such as caste, ownership of agricultural assets and non-crop incomes. These economic variables, in addition to farm size, may be taken as a proxy for security against risks of a new technology.

Impacts of caste on tree planting   'Lower' castes are defined as those entitled to affirmative action in government schemes, and include the scheduled castes, scheduled tribes, and backward castes. The rest are called 'upper' castes. Numbers of trees planted by the caste categories are given in Table 9.5, showing that among both large and small farmers, more trees were planted by upper-caste households. The average number of trees for an upper-caste large farmer was more than 12 times that planted by a lower-caste small farmer.

Tree planting among upper-caste households is likely strongly related to their position in the community, and the control this gives them over public resources. Upper-caste farmers are generally better educated and have more relations and friends in the bureaucracy. This gave them better access to seedlings, markets and extension advice. They also often have a strong

Table 9.5 Trees planted by different caste groups

	Average number of trees planted by		
Farm size class	Upper castes	Lower castes	All households
Large (> 2.5 ha)	521	331	471
Small (0.5–2.5 ha)	115	43	75
All households	324	112	231

influence in village institutions, such as village councils and cooperatives, giving them better control over scarce public resources, such as subsidized seedlings from government depots.

During the colonial period, upper-caste households were often non-cultivating landlords. They also often have a culturally derived aversion to manual labour, because of caste restrictions. Tree growing reduced the need for them to have to rely on tenancy arrangements to keep their land cultivated on the one hand, while suiting their cultural attitudes toward manual labour on the other. Even during the colonial period, upper-caste households had a tradition of planting fruit trees on grove lands as a labour-extensive land-use strategy. The data showed a positive correlation between those who had grove or fallow land and the numbers of trees planted.

Agricultural assets and non-crop incomes Table 9.6 considers the association between trees planted, agricultural assets, and non-agricultural incomes. Farmers who planted more than 100 trees had greater assets and more diversified sources of income, than farmers who had planted fewer trees.

Table 9.6 Agricultural assets and non-crop incomes

Numbers of trees planted	Numbers of households	Agricultural assets, in Rs	Non-crop incomes, in Rs
less than 10	26	22 975	3 791
from 11 to 100	66	29 506	5 806
from 101 to 500	29	48 223	6 913
more than 500	25	70 535	23 572
total/average	146	39 086	8 709

Notes: Tree values are not included as agricultural assets. Income from the sale of trees is not included in non-crop income.

Differences in non-crop incomes across different groups were particularly striking for those which planted more than 500 trees. Among small farmers, non-crop incomes of those planting more than 500 trees were nearly three times the non-crop incomes of non-planters. Among all farmers, tree planters had non-crop incomes which were more than six times greater than non-planters.

With larger non-crop incomes, tree planters were far less dependent on the income lost by diverting agricultural land to trees and could afford to be less risk averse than other households. With greater non-farm income generating opportunities, tree planting households in some respects encountered greater labour constraints in the first place, as household labour was less able to be involved in farm work or labour supervision.

Summary and discussion

In summary, field studies in Uttar Pradesh have shown that households which planted eucalypts in this agriculturally prosperous region of north-western India were more likely to be absentee landowners and large farmers, with the support of substantial non-crop incomes. They were more likely to belong to 'upper' castes, and possessed more agricultural assets than those who did not plant or planted only a few trees.

This study also shows that, although the planting of trees on field boundaries was common among farmers in western Uttar Pradesh, the vast majority of planted farm trees were to be found in woodlots, replacing annual crops. Woodlot establishment was more common amongst the most supervision-constrained households, which also had to hire more labour to cultivate and harvest their annual crops.

In contrast, in the more subsistence-oriented villages of eastern Uttar Pradesh, large landowners (with greater access to cheap labour or with docile tenants) had less of a compulsion to change their cropping patterns, or tenancy arrangements in favour of labour extensive crops such as eucalyptus. Bardhan (1979) has argued that labour-intensive technical change is likely to result in a higher incidence of tenancy. If one accepts the converse, our data support this view: in no instance were trees found to be grown under a tenancy arrangement.

In the commercialized north-western region of India, rates of adoption of high yielding crop varieties among small farmers were similar to adoption rates among larger farmers (the first adopters) within a few years after the introduction of these innovations (Prahladachar 1983; Lipton and Longhurst 1989). Improved varieties offered high yields with low inputs. The good infrastructure in the north-west meant that any farmer could readily purchase the required inputs in whatever quantities were needed. Small farmers could see the results on their own fields in a short period of a few months, and gain confidence.

This was not the case for eucalypts. Small farmers could be handicapped in respect of information, education, and access to government officials, attributes often correlated with caste. Practically, this meant they had less access to subsidized seedlings and technical advice about tree growing. Eucalypts take at least 6 years to mature, and even on the fields of large farmers, trees were not able to provide a quick demonstration of productivity or financial viability. The ownership of productive assets such as milch-cattle or tractors, as well as diversified sources of non-crop income, also gave farmers greater control over their production and marketing environment, enabling them to risk adopting a new crop like eucalypts.

Fewer labour supervision constraints, coupled with less access to inputs and technical advice, the long time to maturity, and less access to non-crop income all operated against the adoption of eucalypts as a cash crop by small farmers. These farmers are almost entirely dependent on the land, especially in a region of fertile soils where annual crops offer a stable income.

A review of tree planting practices across several continents concluded that large farmers often grow trees to meet market demands, while small farmers have greater constraints in this respect (Food and Agriculture Organization of the United Nations (FAO) 1986). Ellis (1988) holds that small farmers, because of considerations of family security, allocate a higher proportion of their land to subsistence food crops. Arnold (1987) argues that under particular conditions, as farm size declines, production objectives can change from the growing of food to the generation of income, with which small farmers can buy food, and that the incorporation of trees into farming systems is often a response to this change in production objectives.

This study suggests that the situation in western Uttar Pradesh is more complex. Small farmers and large farmers followed similar cropping patterns, used similar amounts of chemical fertilizer, and sold a substantial part of their produce in the market. What small farmers lacked was sufficient surplus and staying power, which would have enabled them to tide over the period until maturity, during which trees yielded no income. The distinction is perhaps better understood not in terms of the advantage of subsistence crops over cash crops, but in terms of the advantages of short-duration over long-duration crops. In this case, tree planting is clearly an outcome of the processes of higher risk bearing capacity and multiple sources of income associated with large holders.

9.4 Implications for policy and planning

The two studies reviewed here suggest that an assessment of the impacts of tree planting on employment must consider how different groups of households are affected, depending on their access to sources of labour (both from the household and from labour markets), returns to different land uses at different levels of labour utilization *vis-à-vis* other opportunities for income generation, access to capital to finance the hiring-in of labour, the operation of rural land markets, and the proximity of the household to other labour markets. Land-use choices reflect different levels of labour needed to make land productive, as well as markets for commodities and inputs which define the margins which households can hope to earn.

An additional consideration is the *quality* of agricultural land *within* a holding as well as across holdings in a particular agroecological zone. Land resources are seldom homogeneous. Some holdings or bits of a holding may be of better or worse quality for specific crops, and the use of certain parts of a holding for tree growing may reflect this. Particularly in hilly areas, the decision to use different parts of the holding for different crops may be a function of the household's view that some crops may be more difficult to manage if they are planted on steep slopes. Similarly, households which are endowed with land resources of a particular quality may be more inclined to adopt tree planting than others.

As with a household's land resources, labour resources are seldom homogeneous across households. Households with a larger number of residents may have readier access to household labour than smaller households. Particularly in Africa, polygamous households tend to have more land under cultivation, support a greater number of residents, and are, generally, wealthier, and these factors, in turn, greatly influence the choice of land uses.

Household labour supplies and the number of individuals resident on a farm may be a function of the household's age structure. It is often characteristic of older households, for instance, that they support a smaller number of residents because children have moved away to their own farms or to find work in other labour markets. Older households may have less of a *need* to cultivate their holding intensively because their capital requirements are lower, their savings may be enough to support them, or because wage remittances from non-resident children provide them with an adequate income. Access to labour, then, may be related to quite specific household characteristics, and when labour is constrained by any of these characteristics, there may be a greater incentive to use the holding for cultivating crops, such as trees, which have fewer labour requirements.

In perfectly operating markets, the use of land for growing trees would be less likely if 'surplus' land (defined as the land which a household is unable or unwilling to cultivate intensively because of capital or labour constraints, or because productive demands are lower) would be sold or rented out. In practice, intergenerational concerns may limit the household's willingness to sell land which would otherwise be passed on to the next generation, and tree planting may be seen as a low cost, labour extensive means of keeping land productive.

There may be constraints to renting surplus land out, because rental markets may be poorly developed or because security of tenure to the original right holder might be jeopardized. A household which rents land in must expect to be able to generate enough of an income to cover its rental costs. Households which rent land in are often only able to cultivate annual crops, because the establishment and cultivation of more lucrative permanent cash crops might imply some sort of longer term rights of tenure to rented-in land. All of these features of land use and tenure can work in ways which favour the planting of trees on land which households are otherwise unable to use, instead of renting it out or selling it.

We have attempted to show that supplies and demands for land and labour interact in sophisticated sorts of ways, which sometimes influence the decision of households to cultivate and manage trees on their holdings. In summary:

- Tree planting may be seen to be a feasible land-use option when the opportunity costs of using household labour on-farm are high because there are good wage opportunities in other labour markets.
- Where household labour resources are constrained, problems with supervising and hiring-in labour to more intensively cultivate annual crops can act as incentives for households to plant or to maintain trees instead of other crops.
- Labour-to-land ratios are often determined by demographic changes in household composition, with older households having a smaller resident labour force on which to draw. The outcome can be the adoption of less labour intensive forms of land use, such as tree growing.
- The need to intensively cultivate a holding may be less among older households with fewer residents, which may instead be more dependent on savings or on wage remittances from non-resident children for their support. Trees may be planted in order to maintain a productive crop on land which will eventually comprise the household's inheritance.
- The quality of land within a holding, as well as across holdings in a given agroecological zone, may be highly heterogeneous. This, in turn, may mean that some holdings, or some parts of some holdings, require much more labour to cultivate than others. Trees may be planted in these areas in order to even out labour demands.
- The sale of land surplus to the household's immediate needs may be undesirable because of intergenerational concerns, and the need to retain resources which can be passed on to the next generation. The renting out of surplus land may be undesirable because it might jeopardize the tenure holder's long-term rights of ownership. In these circumstances, trees may be planted and maintained, as an alternative to letting the land lie fallow.

These interactions are often wholly exogenous to rural afforestation policy and planning processes. Consequently, Forest Departments and other agencies with rural forestry responsibilities are often entirely unable to influence the most signficant factors which bring about the adoption of tree planting practices on farms.

Notes

1. A review of studies which have investigated the economics of agroforestry systems concluded that household or farm-based studies were indeed rare. Only 20 per cent of the 230 studies reviewed used data collected from on-farm research. Many studies depended at least partly on the results from on-station trials (28 per cent) or on estimates (43 per cent). Analyses which were intended to evaluate the post-project impact of agroforestry investments accounted for only 13 per cent of the studies reviewed (Swinkels and Scherr 1991).
2. For the purposes of this discussion, *holdings* are made up of one or more geographically distinct *parcels*. *Parcels* are made up of one or more adjoining *plots*.
3. Kenya National Archives AGR/4/343, 10 April 1941. *Wattle, General 1937–1957.* 'Should black wattle trees be grown in Kikuyu Country?' Drafted as an article by the Senior Agricultural Officer for Central Province, at the request of the Director of Agriculture, for publication in the Swahili language newspaper, *Baraza*.
4. This is particularly the case where the opportunity cost of land is high. The previous chapter, for instance, suggested that large numbers of small farmers planted trees instead of other cash crops in anticipation of good market opportunities and because of the perception that markets for wood products were less risky than other markets. The risk of planting trees in these circumstances is much less than when alternative land uses are assuredly lucrative. This chapter specifically discusses the case of Uttar Pradesh, where the green revolution greatly increased the opportunity cost of land, and where the risks, consequently, of tree planting were much higher than they were in areas where the opportunity cost of land, in the face of lucrative wood market opportunities, was lower (as in the previous chapter).

References

Arnold, J.E.M. (1987). Economic considerations in agroforestry. In *Agroforestry: a decade of development* (ed. H.A. Steppler and P.K. Nair), pp. 173–90. International Council for Research in Agroforestry, Nairobi.

Bardhan, P. (1979). Class relations in Indian agriculture: a comment. *Economic and Political Weekly*, **14**, 857–60.

Bardhan, P. and Rudra, A. (1978). Interlinkage of land, labour, and credit relations: an analysis of village survey data in east India. *Economic and Political Weekly*, **13**, (6/7), 367–84.

Basant, R. (1987). Agricultural technology and employment in India: a survey of recent research. *Economic and Political Weekly*, **22**, 1348–64.

Beteille, A. (1969). *Caste, class and power.* Oxford University Press, Bombay.

Byres, T.J. (1981). The new technology, class formation and class action in the Indian countryside. *Journal of Peasant Studies*, **8**, 405–54.

Chambers, R.J.H. (1988). *Trees as savings and security for the rural poor*, Gatekeeper Series Paper No. 3. International Institute for Environment and Development, London.

Clayton, E.S. (1961). Economic and technical optima in peasant agriculture. *Journal of Agricultural Economics*, **14**, 337–47.

Collier, P. and Lal, D. (1986). *Labour and poverty in Kenya, 1900–1980.* Clarendon Press, Oxford.

Cowen, M.P. (1978). Capital and household production: the case of wattle in

Kenya's Central Province, 1903–1964. Unpublished D.Phil. dissertation. University of Cambridge.

Dewees, P.A. (1991). The impact of capital and labour availabilty on smallholder tree growing in Kenya. Unpublished D. Phil. thesis. University of Oxford.

Ellis, F. (1988). *Peasant economics: farm households and agrarian development.* Cambridge University Press, Cambridge.

FAO (1986). *Tree growing by rural people*, FAO Forestry Paper No. 64, Food and Agriculture Organization of the United Nations, Rome.

Gregersen, H., Draper, S. and Elz, D. (ed.) (1989). *People and trees: the role of social forestry in sustainable development.* Economic Development Institute of the World Bank, Washington, DC.

Harriss, B. (1972). Innovation adoption in Indian agriculture: the high yielding varieties programme. *Modern Asian Studies*, **6**, 71–98.

Jose, A.V. (1988). Agricultural wages in India. *Economic and Political Weekly*, **23**, A. 46–58.

Leach, G. and Mearns, R. (1988). *Beyond the woodfuel crisis: people, land and trees in Africa.* Earthscan, London.

Lipton, M. (1985). *Land assets and rural poverty*, Staff Working Paper No. 744. World Bank, Washington, DC.

Lipton, M. and Longhurst, R. (1989). *New seeds and poor people.* Unwin Hyman, London.

Patnaik, U. (1976). Class differentiation within the peasantry: an approach to the analysis of Indian agriculture. *Economic and Political Weekly*, **11**, A. 82–101.

Patnaik, U. (1986). The agrarian question and the development of capitalism in India. *Economic and Political Weekly*, **18**, 781–93.

Prahladachar, M. (1983). Income distribution effects of the green revolution in India: a review of empirical evidence. *World Development*, **11**, 927–44.

Rutten, M. (1986). Social profile of agricultural entrepreneurs: economic behaviour and life-style of middle-large farmers in central Gujarat. *Economic and Political Weekly*, **21**, A. 15–23.

Srivastava, R. (1989). Tenancy contracts during transition. *Journal of Peasant Studies*, **16**, 339–95.

Swinkels, R.A. and Scherr, S.A. (1991). *Economic analysis of agroforestry technologies: an annotated bibliography.* International Council for Research in Agroforestry, Nairobi.

Part IV
Conclusions

10 Retrospect and prospect

J.E. Michael Arnold

10.1 Introduction

In this final chapter we examine the extent the information deployed in the earlier parts of the book permits a fuller understanding of the circumstances in which tree management forms part of farmer strategies. In particular, we consider whether the patterns of tree growing behaviour that have been observed and analysed enable the formulation of stronger hypotheses about the role of trees and tree products, and about requisites for tree growing by farmers, and provide a more focused starting point for the process of defining the need and potential for policy, programme or project interventions.

As could be expected from such a complex and multi-faceted subject, greater knowledge does not necessarily lead to more clearcut prescriptions. Some of the earlier assumptions have been shown to be unsound or incomplete, but the old theories cannot be replaced by equally simple new orthodoxies. As the variety and complexity of the array of different roles trees play in rural household livelihood strategies becomes clearer, it is evident that there are few answers of general application. Nevertheless, a number of issues of wide application can be identified.

10.2 Trees and the dynamics of rural change

Processes of change

As has been demonstrated in nearly all of the situations examined here there have been, and continue to be, significant changes in the way farm households employ trees and tree products over time, with many of the changes being both substantial and rapid. Furthermore, the general trend in different regions experiencing agricultural intensification has been towards more intensive tree planting.

The 'forestry' and 'welfare' approaches to analysis of the place of farm trees that predominated during the 1970s and 1980s have proved to be of only limited value in understanding this phenomenon, or in identifying and defining needs for intervention (see Chapter 1). With their restricted focus on particular needs and products, they tended to obscure the dynamics of farmers' economic responses to changes in demand and supply and to scarcity and abundance. Production and use of tree products at the village level are in practice often embedded in complex resource and social systems, within which most of the factors that affect our ability to intervene with forestry solutions are of a non-forestry nature. They are primarily human factors, connected with the ways people organize the use of their land and other resources. They are therefore unlikely to be successfully understood through approaches that address only a single element of the situation. The discussion in Chapter 7 has shown that lack of success of many early tree growing projects was largely because they took shape as a response to an energy supply problem, rather than as a response to local needs for trees and tree products. Because little was known about how farmers respond spontaneously to declining supplies of fuelwood, the case for tree growing was not balanced against alternative courses of action.

To pursue tree growing by farm households as primarily a vehicle to combat deforestation, or for poverty alleviation, is likely to be counterproductive. It is necessary to identify those issues that tree growing can address, and those that it cannot, and these will vary greatly from situation to situation. Where trees do serve a purpose in terms of the household objectives, they will often contribute to meeting environmental and distributional concerns as well. But it needs to be clear that any broader resource and environmental benefits that may accrue from farm tree stocks emerge as a by-product of, and are subordinate to, farmers' pursuit of their livelihood goals.

Using the concepts of livelihood security and induced innovation as a starting point, tree planting can be explained as being one or more of the four categories of response to dynamic change that were identified in the introductory chapter:

- to maintain supplies of tree products as production from off-farm tree stocks declines due to deforestation or loss of access;
- to meet growing demands for tree products as populations grow, new uses for tree outputs emerge, or external markets develop;
- to help maintain agricultural productivity in face of declining soil quality or increasing damage from exposure to sun, wind or water runoff;
- to contribute to risk reduction and risk management in face of needs to secure rights of tenure and use, to even out peaks and troughs in seasonal flows of produce and income, and in seasonal demands on labour, or to provide a reserve of biomass products and capital available for use as a buffer in times of stress or emergency.

With most situations experiencing reduction in access to off-farm supplies, growing demand, declining site productivity and increased exposure to risk, it should not be surprising that tree planting activity does increase as agriculture and land use become more intensive. Though it would be incorrect to assume that this always happens, there is a general progression towards more *planted* trees as agriculture and pressures on land intensify, and existing tree stocks diminish, within most systems.

It is also clear that a substantial part of the increase in farm tree growing reflects the fact that, as pressures on their labour and other resources increase, many farmers are responding by *reducing* the intensity of use of land, or part of their land. Faced with shortages of labour and other inputs to agriculture (compost, fertilizer, etc.), farmers abandon their poorer or more distant lands, or put them under less intensive use, in favour of concentrating the use of available inputs on the more favourable sites.* By no means all agricultural intensification results in increased competition for use of the land.

The analysis of the potential for change therefore needs to be set within a framework that reflects these features. It is also important to recognize the historical context. Present changes are often just the latest stage in a process of co-evolution of society and environment that has been taking place for millennia.

The shift of tree resources from public to private property

At its simplest the shift from public to private resources implies that, as access to particular trees of value off-farm declines, stocks of these species are established on-farm. However, some of the changes that result in depletion of common pool resources, and undermine their collective management, may also alter demand for the tree products previously supplied from these off-farm resources. Improved access to markets, for instance, could make available fertilizer in place of green mulch, and create outlets for livestock products that cause livestock management to shift towards pasture crops and stall-feeding rather than grazing. It cannot be assumed, therefore, that the tree products most in demand from off-farm sources will necessarily determine what trees farmers wish to establish on-farm.

In addition, in many situations farm households draw on the former for supplies of produce that could not be produced on-farm. In the drier areas of India, for instance, the numbers of draught animals needed in order to cultivate enough cropland can only be maintained if there is access to large areas of grazing or fodder off-farm. For other than the largest farms it would not be possible to establish sufficient tree stocks on-farm to maintain livestock in these numbers. As off-farm resources of some tree products decline, users may thus be forced to reduce consumption of tree products or to adopt options that do not depend on access to the latter. For the wealthier, this may take the form of investing in alternatives such as tractors rather than draught animals. For the poorer, there may be no viable option that would keep the farm system functioning in its present form once access to off-farm supplies is lost, or is reduced below a critical level.

* See, for example, Tamang (1993) on Nepal, and Lagemann (1977) on Nigeria.

This suggests that on-farm tree growing is most likely to form a significant component of the farm household system in higher rainfall, more arable, areas that are less dependent on large quantities of tree biomass and where higher site productivity, and freedom from free-ranging livestock, enable needed tree products to be produced efficiently on-farm. This pattern is borne out by the study evidence from eastern Africa that was presented in Chapter 5, which showed that planted trees figured most prominently in the more arable high rainfall zones.

In dry regions, with extensive agricultural and livestock systems that are dependent on biomass products and woody perennials, but ecological conditions that constrain options for tree growth, woodland management usually offers greater potential than tree planting as a means of increasing supplies of outputs such as fodder and grazing. Even simple management interventions can often produce substantial increases in productivity of such vegetation (e.g., Taylor and Soumaré 1984), but can be inhibited by an array of disincentives that constrain effective collective control and management of natural vegetation, or more intensive management on private land. Many of these constraints relate to issues of control over use of land and of trees on the land.

Changes in tenure and control

In Africa, local control of public forest resources has widely been undermined by the transfer of large areas of land from communal to State control. Management has changed from use-rights based on clan-membership to the exercise of State-granted privileges and management by restriction and exclusion. With government legislation having become necessary for any change to established practice, groups are discouraged from organizing to manage their local resources (Shepherd 1992). Similarly it has been argued that possibly the most important factor in undermining local collective control of common pool resources in India has been the replacement of local leadership and authority with centralized political control—'the ever increasing tendency of the State to expropriate the initiatives and activities which belong to people' (Jodha 1990).

It has long been argued that private growing of such a long-gestation crop as trees will occur only if there is security of tenure over the land on which they are to be planted. However, the thesis that this degree of security can only be provided by private ownership of the land has increasingly been questioned. The occurrence of privately planted trees in a wide variety of tenurial contexts indicates that such generalizations are not necessarily accurate.

In eastern Africa, where customary systems of collective control of the land predominate, with individual households having rights of cultivation and use on an area of land, systems have long been in existence that distinguish security of access to trees from security of access to land (Fortmann 1987;

Warner 1993). Persons who plant trees are assured of continued rights to the produce even after they have relinquished control of the land on which the trees are located. As is demonstrated in Chapter 5, the most important factor affecting tree growing in such systems appears to be the existence or absence of rights of exclusion, in particular, exclusion of grazing on the household's fallow fields. Where this is discouraged, because livestock management is important, or where it cannot be enforced, tree growing is unlikely to take place. Where farmers can exercise this degree of control, economic factors are probably more important than land tenure in determining decisions about tree growing (Cook and Grut 1989; Godoy 1992; Shepherd 1992).

Similarly, tree growing is limited where outsiders have rights to graze on private land. Thus in regions where farmers open their fields to free grazing by others during the post-harvest season, trees are likely to be confined to homestead and other protected areas. In the situation in western Rajasthan that was examined in Chapter 3, it was found that the inability of the individual to capture the benefits of doing so, in the face of seasonal rights of access by others, also discouraged investment in improving the productivity of existing woody vegetation on farmland.

Where land has been reallocated, creating freehold or other individual rights to land in place of collective ownership, changes in tree growing patterns have occurred. However, the changes could as well be explained as responses to decreased access to previously available common property resources, and to other changes in land use, as to increased security of tenure. The more intensive management of *Prosopis cineraria* as an intercrop in western Rajasthan accompanied reduction in access to off-farm sources of arboreal fodder. In parts of central Kenya, increased private tree planting accompanied reduction in access to other supplies, and the necessity to relocate trees from fields to boundaries and homestead areas in order to accommodate the more intensive cultivation practices that became possible with rights of exclusion (Brokensha and Riley 1987; Shepherd 1989).

The need to increase security of tenure or access in order to encourage private tree growing may therefore occur less often than tends to be assumed in government programmes and project design. Indeed, attempts to change tenure can be counterproductive. Past changes have often engendered a strong distrust of government intervention in this area. Moves to alter control of land by creating individual titles to common pool resources can disenfranchise large segments of the local population. The prospect of change can thus itself introduce uncertainty, and so may inhibit investments in long-term activities such as tree growing.

Where the present situation does seem to pose a constraint, it may be more realistic to seek solutions that can be effected within the existing legal and tenurial framework, than to try and alter the latter. Changes in both formal and customary tenure are usually difficult to accomplish, so that it can be unrealistic to design project interventions that require such changes. Lesser changes, such as relaxation or abolition of the regulations that control commercial sales of private tree products in some parts of India (Chapter 8), could often provide the necessary assurance that farmers need.

Nevertheless, there are situations where tenure or control restrictions do determine tree growing decisions. Thus, trees are not found as a sharecropper crop in Pakistan because sharecropping contracts typically run for 1 year, and therefore exclude rights to multi-year crops (Chapter 4). In western Kenya the right to make decisions about trees, and about allocation of income from tree products, is reserved to men; a growing constraint as more farm households are run by women (Chapter 6).

Tree growing can also be shaped by linkages between the presence or absence of trees and control of the land. In some situations legislation or customary practice require that a settler clear land of trees in order to establish rights to the land. In others, rights to use of land may be established by planting trees on the property, or to mark its boundaries. While this may encourage people to plant trees to establish ownership rights, it may result in those who presently control the land prohibiting those who have temporary rights of use of it, such as tenants or squatters, from doing so.

In many countries the State is empowered to appropriate forest or woodland areas. While often intended to bring threatened forests under sound public management, this approach may discourage private tree planting because it introduces uncertainties about rights of ownership and usufruct. In Pakistan, the first question farmers asked about the recent government programme to encourage private tree planting described in Chapter 4 concerned who would own the trees, and would the government take over the land on which the trees would be planted. Where governments intervene to tighten control over forest resources on public land, this can undermine or eliminate local rights of use, and can accelerate the shift towards greater dependence on privately planted resources. The increase in on-farm trees in the hill areas in Nepal discussed in Chapter 2, for instance, started soon after the government nationalized all forests, in the late 1950s.

These linkages with ownership and control distinguish trees from most other land uses and crops. Decisions about tree growing would be facilitated if trees grown as a crop could be more clearly distinguished from issues of control of land. Though some of the practices that give rise to these problems are deeply embedded in customary practice or law, and cannot be readily changed, others often could, notably those arising from uncertainty about, or incorrect application of, government powers with respect to trees and tree-bearing lands.

Agrarian transition and the growth of markets

As agriculture shifts from a predominantly subsistence basis to greater involvement in market transactions, tree growing at the farm level becomes exposed to a number of influences. Markets for factors of production affect the availability and cost of land, labour and capital, and choices between activities that draw upon these factors in different amounts and proportions. Access to purchased inputs such as fertilizer can permit shifts away from extensive land uses involving trees. Access to market outlets for tree products can extend the range of the farm household's income generating options.

A distinction can usefully be made between local rural markets, and the growth of urban and industrial markets for tree products. Local markets for fruits, fuel, poles and other tree products develop, often first as barter trade, as shortages emerge, as increasing demands on the time of women (and other household members) leave less time for gathering what is needed to meet household needs, and as rising cash incomes allow some the option of purchasing rather than gathering or growing. Households that are managing tree stocks in order to provide themselves with such products will sell what is surplus to their needs, or to exploit the opportunity to generate additional income. The study in western Kenya discussed in Chapter 6 found that, in an area where meeting household needs was still the main goal, growing of poles, and to a lesser extent other tree products, had become an important part of overall cash strategies of many households; with sales within communities or between neighbours still predominating.

Participation in urban and industrial markets for wood products is more likely to be practised by farmers in areas where the process of agrarian transition has shifted towards greater involvement in commodity markets and an entrepreneurial approach to agriculture based on cash crops. In these markets, however, farmers can encounter forms of competition and policy constraint that can make it difficult for them to compete. Most urban wood fuel markets are still supplied by mining natural tree stocks, with producers paying little if anything for the raw material, so that the cost of fuelwood delivered to the market consists mainly of transport costs. In most countries, much fuelwood, pole, and other categories of wood, also comes from State forests and plantations, and is sold at administered prices. Prices of fuelwood are frequently subsidized in favour of urban consumers, for the same political reasons that urban food prices are kept artificially low (Chapter 8). Private producers are also frequently subjected to controls on harvesting, transport and sale, designed to protect against illegal felling for sale from State forests. Resulting cumbersome and costly bureaucratic procedures lead to dependence on intermediaries, and often fragmented and inefficient marketing structures.

A combination of these factors helps explain the very limited occurrence of private production of fuelwood and poles for urban and industrial markets. This is illustrated by the comparison of three different situations that appears in Chapter 8. In the Sudan, production by farmers is simply not viable in competition with the low cost supplies from wood generated through agricultural land clearing. In both central Kenya and northern India the cultivation of trees as cash crops did emerge but was curtailed in the face of competitive and policy constraints. Unless there is action to remove such impediments, urban and industrial markets are likely to be less important than local rural markets in most tree growing situations.

10.3 Balancing household needs and market opportunities

Elements of the balance

One of the consequences of the 'basic needs' and 'deforestation' approaches to stimulating and supporting private tree growing by farmers, was a tendency to develop projects as though they were effectively isolated from some of the key influences on them, in particular, economic forces. The assumption that farmers plant trees to meet subsistence or environmental needs, and that these are not bought or sold in the market place, was reflected in projects designed as though they were divorced from and immune to market forces. Some indeed attempted to prevent participants selling their produce on the grounds that this was contrary to the service function assumed to be the goal of farm forestry (Cook and Grut 1989).

Although some of the instances where farm households adopted trees solely or primarily as a cash crop are striking, as in northern India, in most of the situations studied self-sufficiency in particular tree products proved to be the primary objective. However, with the growing dependence of farm households on income to meet at least part of their needs, as forest products such as fuelwood, fodder and fruits become progressively commoditized, and the market-place provides opportunities to substitute purchased inputs such as fertilizer for inputs previously supplied by growing trees, the distinction between production for subsistence or sale has progressively less meaning. Not only will producers sell what is surplus to their subsistence needs, but they will sell a commodity needed in the household if the opportunity cost of doing so is advantageous, hence the widespread phenomenon of households short of fuelwood selling wood. There are therefore seldom clearcut distinctions between the two.

Farmers are likely to enter the market for tree products where they lack other income opportunities, as was the case with black wattle in central Kenya, or where returns from tree crops appeared to be more attractive or stable than from alternative crops, as was the case during the phase when farmers were adopting eucalypts in Uttar Pradesh (Chapter 8). However, farmer decisions about production of tree products for the market will also be influenced by consideration of the role of trees and tree products in meeting objectives other than income generation within their livelihood strategy. These can include the contribution of tree resources to risk management, to optimum use of available land, labour and capital, and to household food security.

Risk management

The use of trees and tree products as a tool in risk management is found to be one of the elements entering most widely into farmer decisions about tree growing (Chambers and Leach 1987). In western Rajasthan farmers cope with an environment subject to repeated periods of low rainfall, and the resulting reduced crop yields, by adopting a strategy of mixing extensive and intensive uses of land through crop–fallow rotation, and the complementary use of annuals and woody perennials, with the latter providing reserves of fodder and other biomass in years of poor rainfall (Chapter 3). In agricultural systems where other forms of accumulating and holding capital, such as livestock herds, are not available, tree stocks often serve this function, to be sold in times of emergency or to meet exceptional financial commitments.

Tree components are also widely included in a farm system because they help even out seasonal peaks and troughs in flows of produce and income and in demands on farm labour. Arboreal fodder, for example, is the principal dry season source of animal feed in both semi-arid areas such as west Rajasthan and the Middle Hills of Nepal. In systems where trees form a substantial part of farm output, as is the case where home gardens are well developed, tree products are also important in diversifying farm output, helping to reduce the exposure of the farm household to failure or price falls on the part of individual staple crops (Ninez 1984). In western Kenya it was found that farmers ensured diversity by adopting several different tree species even for a single use (Chapter 6). Farmers also tend to favour multi-purpose trees with multiple uses, as these provide them with more flexibility in responding to changing household needs and market conditions.

Trees can make other contributions to a more stable and less risk prone system, notably in protecting crop production against damage or deterioration. The farmers reported on in western Kenya, for instance, have recently increased the growing of trees on-farm in order to provide windbreaks and to produce green manure. Farmers will use trees that complement or supplement existing crops in different ways and locations to trees that compete with the latter. Farmer efforts to maintain and intensify cultivation of *Prosopis cineraria* as an intercrop in Rajasthan reflected its value in enriching the soil and protecting the adjacent pearl millet against wind, as well as its value as a fodder crop. In contrast, the costs of crop losses in the vicinity of the eucalypts they had planted was a factor in the decision of farmers in Uttar Pradesh to discontinue cultivation of the latter.

The importance of risk management in poor farmers' strategies has implications for analysis of the role of on-farm trees. It is likely that farmers use lower implicit discount rates in making decisions about activities that can contribute to risk minimization, or to meeting self-sufficiency needs, than they do about income generating activities. Low yielding long gestation tree options that appear to produce low financial returns may be seen to be viable components of the system once their role in relation to risk or food production is understood.

Resource availability and allocation

The progressive reduction in farm size, and deterioration in productivity of farmland, that accompany the pressure of rising rural populations on the arable land base, mean that many farmers are confronted with limited land resources. As a crop that produces low returns per unit of area, trees generally become restricted to homesteads, boundaries and other niches where they do not compete with the agricultural crops as land becomes the limiting factor. However, there can be exceptions to this where tree outputs complement or supplement crop outputs; thereby increasing total returns per unit of area. Home gardens, with their vertically layered structure of trees, shrubs and ground cover crops making effective use of space above and below the soil surface, provide notable examples of this. Alley cropping, involving cultivation of hedges to provide green mulch where capital scarcity restricts the use of fertilizer, is another labour-intensive application of woody perennials employed to maintain the productivity of land-constrained farms.

As farm households in many situations increasingly have to depend on income earned from employment off-farm, the amount of labour available for farm work is progressively becoming the main resource limitation determining farmer options. The fact that trees need low inputs of labour to establish, and even less labour to maintain, helps explain some of the increase in on-farm tree growing. As is illustrated by the information in Chapter 9, the use of trees as a response to labour becoming a limiting factor takes a number of forms. In central Kenya it was households with fewest resident adult males that were found to be most likely to retain or establish black wattle woodlots, in preference to using the sites for the growing of the much more labour and capital intensive crop of tea. In Uttar Pradesh it was the more asset-rich households, with significant off-farm income, and seeking to minimize labour supervision requirements, that were most likely to grow eucalypts as a crop.

The last of these characteristics demonstrates another attraction of trees to some: their suitability as a crop for absentee owners or farmers, as an alternative to leasing the land out with the risk that entails of losing control over it, or as a way to establish continuing rights to idle land. In Nepal, the advantages of having tree stocks closer to the users, as the amount of labour available to gather fodder, mulch, livestock bedding and fuelwood declined, appeared to be an additional factor in increased tree growing (Chapter 2).

Household food security

As livelihood strategies have increasingly focused on income generation in order to ensure access to sufficient food, and consequently on cash crops and off-farm employment, tree crops have emerged in some situations as a cash crop that matches farmers' changing needs and possibilities. In addition to being a crop that can be cultivated with low inputs of labour, trees require only limited capital to establish and maintain. But other features of some trees may limit their suitability as cash crops.

Cash crops, in general, may undermine the household's capacity to meet its essential food needs in a number of ways. Food prices may rise locally because of the transfer of land out of food production, or because of costly transport and marketing. A drop in cash crop prices will reduce income with which to purchase food; a danger that is accentuated the narrower the range of cash crops and market outlets the farmer is dependent upon. Cash crop income is likely to be lumpy in its availability, leading to periods with little income, and the shift to cash crops may reduce employment opportunities. With rising incomes, spending behaviour is likely to change away from foods, and within foods away from staple foods; a trend more likely to happen if the shift to cash cropping lessens women's control over household resources. Reduction in the area of land available for household production of staple foods, or increased demands on their time because of the cash cropping, can put pressures on their staple food supplies (Longhurst 1987a).

The suitability of crops of trees such as eucalypts can be questioned on some of these counts. They take more than one season to mature. Research, education and marketing services related to tree crops are concentrated on male farmers, who may also control access to the income from tree products. Producing only a single product, eucalypt crops are potentially vulnerable to market fluctuations, and thus to income fluctuations, and they provide income in 'lumps' followed by periods of several years with little or no income (Longhurst 1987b). However, this needs to be set against the attribute of being able to realize income from trees at a timing of the farmer's choosing; providing a flexible form of capital that can be drawn upon to help meet contingencies or to finance periodic lump sum outlays, such as weddings or the purchase of land (Chambers and Leach 1987).

As they require less labour to establish and maintain, growing of trees can transfer land use from other crops with net loss of employment. The concern that has been expressed, notably in India, that the expansion of the growing of eucalypts as field cash crops is reducing rural employment, and has been diverting land from production of essential foods, therefore merits attention. However, it tends to overlook factors that are causing farmers to withdraw land from coarse grain production, such as poor and unstable grain prices and rising grain cultivation costs, and to find more stable and remunerative crops and less costly forms of land use, and the features of tree growing that can make it a logical response to these changes. Such shifts in land use in favour of tree crops therefore reflect developments encouraging farmers to reduce the intensity of use of some categories of their land. To the extent that these trends are deemed undesirable, they are more likely to be remedied by policy changes that influence input and output prices for alternative agricultural crops, than by imposing regulatory restrictions on the cultivation of trees.

The multi-purpose trees and multi-species systems that are found in most farming situations, on the other hand, are more likely to contribute to a sound mixed subsistence/cash crop household economy. Priority nearly everywhere is given to establishing fruit trees. Fruit or fodder products, and shade, protection, green manure or soil amelioration from the presence of trees being grown as cash crops, all contribute to staple food supplies, as well as to income generation. Virtually all trees will also produce some fuelwood for use in the household. The counterseasonal evening out of income and labour use that tree crops commonly permit is a further positive impact. Tree species that can be intercropped with early yielding crops reduce the problems of delayed returns usually associated with longer rotation tree production. Incremental adaptive changes of this kind are likely to pose less of a risk to small farmers than changes that involve major alterations to existing land use or product flows.

These features help explain the limited occurrence of trees as field crops, and the prevalence of land use patterns in which a variety of trees are fitted into different niches in the farm landscape. It seems likely that the conclusion that can be drawn from the studies in central Kenya and Uttar Pradesh, that the growing of trees as field cash crops to produce wood is likely to be an appropriate option only if the household has access to other sources of income or food, and if there are reasonably secure markets for the tree products, is of quite widespread application. However, markets for tree products can be important to the poor if the costs and risks involved are low, e.g. low costs of entry, early returns, market channels that serve small-scale as well as large-scale producers, and production systems that can be developed incrementally and do not put other parts of the farm system at risk. For example, farmers in the area studied in western Kenya, who had been increasing tree growing to meet their household needs for poles, fruit and fuelwood, expanded their production of these products when local markets for them developed, in order to sell some part of their output (Chapter 6).

10.4 Policy and research implications

Defining the scope of interventions

Policy and project interventions in support of tree growing by farmers have often been poorly matched to the role of trees and tree products. High priority still needs be given to improving understanding which problems can be addressed through tree growing, and which cannot. Trees and tree products can contribute to meeting a number of household welfare goals, both directly as a source of food, fuel, housing materials, farm inputs, etc., and indirectly through their impact on maintaining site productivity, and in reducing risk. However, in a given situation the potentials of tree-based interventions need to be compared with that of alternative ways of achieving the same goals.

Similarly, a clearer understanding is needed of whether there are market or government failures that constrain or distort the present situation, and if so whether and how they might be remedied or alleviated through intervention. Too many of the interventions to date have sought to encourage tree growing where trees are not an appropriate component of the farm household economy, or have attempted to induce growing of inappropriate trees. Others have pursued solutions that would require changes in the institutional or social framework that could not realistically be achieved in connection with tree growing, or have failed to focus on the critical areas where change could be brought about.

Improving the range and quality of technical options

The perception that the environmental and fuel shortage problems that were to be tackled through farm forestry needed urgent action, on a massive scale, often resulted in pressures being placed upon forest services to achieve over-ambitious planting or seedling distribution targets—pressures that all too often resulted in priority being accorded to quantity rather than quality or appropriateness (Arnold 1992). Many tree planting support programmes have consequently been characterized by poor technical prescriptions and practices; with implementation being forced to run ahead of capability to provide adequate extension and technical packages, and without sufficient regard being paid to institutional constraints and possibilities.

Though nominally designed to service a 'needs' approach, technical options made available to farmers have seldom been systematically selected to match those needs. A recent review study on the subject commented on the

. . . *appalling casualness* with which the whole question of species choice is approached by the majority of tree planting projects. The project literature evinces little or no awareness that there are different kinds of *tree users* and that the purposes for which trees are planted might vary not only with the type of tree but also with the type of user. In place of a systematic approach to species selection what the project documentation reveals is a general tendency to promote undifferentiated 'tree planting'—*as if all trees were the same!* (Raintree 1991, emphasis as in the original).

One reflection of this has been the bias in choice of trees for projects in favour of a small number of forestry species better suited for production of timber than of the produce farmers seek to obtain from trees.

The study in western Kenya (Chapter 6) showed that, where they did have access to a wide range of tree-based options, farmers can employ a wide range of species, in a variety of different roles and niches, as they intensify land use. This suggests a much broader based approach to extension than has usually been adopted, with farmers being able to choose from a menu of options, in recognition that there is a high degree of variability between, and even within, households for specific tree products and services. This would also increase the likelihood that technical packages are adopted that are compatible with the institutional and policy realities of the situation within which they are to be implemented.

Progress in developing better focused species choices is hampered by the limited amount of applied research that has been done on farm trees, and on competitive and complementarity relationships within intercropping systems. Nor has much of such research results as have been developed been tested on-farm yet. Thus, even with alley cropping, which has benefited from one of the most intensive and thorough research efforts of any innovation in the field of agroforestry, it is still unclear to what extent farmers will find it appropriate (Raintree 1989).

Shifting the focus of policy interventions

One result of the early focus on promoting tree growing as a response to perceived subsistence needs has been failure to match project production to market possibilities, or to link producers to markets. The collapse in pole wood prices in north-west India as large quantities of farmer grown material entered the market in the late 1980s, as a consequence of social forestry support programmes, in part reflected lack of market information, and a lack of attention to the functioning of this emerging market (Chapter 8).

Project interventions have instead centred on provision of subsidized planting stock and/or cash payments to offset establishment and maintenance costs. It is difficult to identify evidence that indicates that this type of support is needed, or that it is effective in increasing farmers' profits. As a crop requiring only low inputs of capital it is not clear that cost constrains many farmers who wish to do so from growing trees. Indirect evidence of the growth of market transactions in seedlings where tree growing has proven to be viable reinforces this view. However, provision of seedlings of species that are not locally available, or that are difficult to raise, can often be valuable.

Recent evaluations of projects in India suggest that there is a danger that intervention in the form of cash subsidies is encouraging tree crops in situations where they are unlikely to be viable or appropriate. In Bihar, farmers appeared to be planting in response to the short-term returns from the cash payments provided rather than the longer term returns from investment in trees, leading to undesirable distortions in land use such as displacement of sharecroppers, and reduction in small farmer subsistence production of food crops to the point where household food security could be adversely affected (Arnold *et al.* 1990). There is also widespread evidence that both seedling distribution and cash subsidies tend to be targeted towards larger farmers, not least because this enables the Forest Service to reach its targets quickly and with the minimum number of transactions.

A more productive approach to identifying policy interventions would appear to be to focus on removing barriers to tree cultivation. Where costs of establishment or husbandry are constraints, more attention could be paid to credit, and measures to reduce costs such as staggered planting, rather than subsidies. It appears, however, that the impediments are mainly on the market and demand side, rather than on the supply side. More attention needs to be paid to policies and practices that presently constrain farmers' access to markets, and that depress market prices for their tree products. These commonly include lack of market information, poorly functioning trading systems serving small producers, competition from subsidized supplies from State forests and plantations, fuelwood prices depressed by subsidies to alternative fuels, and restrictions on private harvesting and trading of wood products (in order to guard against illegal cutting from public forests). There is a danger that by hindering farmer access to tree product markets, governments may inadvertently be interfering with the shift from a subsistence to a market economy.

Consideration of the policy framework influencing tree growing also needs to take account of the impact of policies in related areas. These include policies that affect prices and decisions about alternative crops and land uses, policies that influence adoption of new agricultural technologies such as use of tractors, and policies that affect changes in livestock management and hence demand for grazing and fodder. But there are many others as well that can affect tree management including infrastructure policies that increase access to markets, policies that bear on the functioning of factor markets, and land use policies concerning privatization and control of land.

Improving the information base

By comparison with most other aspects of developing country agriculture, the information base on trees within farming systems that is needed in order to identify needs for interventions, and to design them, is still woefully weak. Little is known about economic responses or even about the physical production functions involved in many agroforestry systems. The historical data on changes in production and use that normally provide the starting point for policy analysis have seldom been assembled for tree resources within agricultural systems. Because of the shortage of detailed 'case' studies of tree management it is seldom possible to examine the likely patterns of change through comparative studies across different situations. Nor are existing secondary data usually spatially organized in a manner that would facilitate such exploration of patterns of behaviour and change. The problem is compounded by the need to look at change over the relatively long time frames associated with production of many tree products.

The approaches adopted in the studies reported on in this book indicate a number of ways in which existing information can be used in order to improve the information base. Archival research can often yield important pointers to past change in the presence of trees within land and resource use, and the reasons for change. Aerial photographic coverage and satellite imagery from different periods can provide more direct and detailed evidence of the nature and extent of past changes in tree cover (and provide a basis for designing follow-up field studies). Secondary data may be reorganized in ways that permit comparison of patterns of tree occurrence and management across different agroecological regions, land use systems, and conditions of wealth and market access. Careful monitoring and evaluation of projects, and of experiments, is another valuable source of information of use in analysis. There are also other techniques, such as landscape modelling, that might be used.

The improvement of data at the policy and planning level needs to be matched by information that improves understanding of the role of trees and tree products in the household economy at the local level. Traditional household and regional surveys have an important role to play in this connection, as do forms of appraisal that involve local people more directly in the planning and decision process. As has been stated in Chapter 6, 'at the local level, participatory planning processes whereby local people and communities can identify their own priorities for tree-growing will probably continue to generate more reliable project agendas than top-down planning'.

References

Arnold, J.E.M. (1992). *Community forestry: ten years in review*, Community Forestry Note No. 7 (revised edition). FAO, Rome.

Arnold, J.E.M., Alsop, R. and Bergman, A. (1990). *Evaluation of the SIDA supported Bihar social forestry project for Chotanagpur and Santhal Parganas, India*. Swedish International Development Authority, Stockholm.

Brokensha, D. and Riley, B.W. (1987). Privatization of land and tree planting in Mbeere, Kenya. In *Land, trees and tenure* (ed. J.B. Raintree), Proceedings of an International Workshop on Tenure Issues in Agroforestry, Nairobi, 27–31 May, 1985. ICRAF and the Land Tenure Center, Nairobi and Madison.

Chambers, R. and Leach, M. (1987). *Trees to meet contingencies: savings and security for the rural poor*, IDS Discussion Paper No. 228. Institute of Development Studies at the University of Sussex

Cook, C.C. and Grut, M. (1989). *Agroforestry in sub-saharan Africa: a farmer's perspective*, Technical Paper No. 42. World Bank, Washington, DC.

Fortmann, L. (1984). The tree tenure factor in agroforestry with particular reference to Africa. *Agroforestry Systems*, **2**, 231–48.

Godoy, R.A. (1992). Determinants of smallholder commercial tree cultivation. *World Development*, **20**, 713–25.

Jodha, N.S. (1990). Rural common property resources: contributions and crisis. *Economic and Political Weekly, Quarterly Review of Agriculture*, **25**, A. 98–104.

Lagemann, J. (1977). *Traditional farming systems in Eastern Nigeria: an analysis of reaction to increasing population pressure*. Africa Studien, Weltforum Verlag, Munich.

Longhurst, R. (1987a). *Cash crops, household food security and nutrition*. Paper presented at the Cash Crops Workshop, Institute of Development Studies at the University of Sussex, January 1987.

Longhurst, R. (1987b). *Household food security, tree planting and the poor: the case of Gujarat*. Paper presented at an IDS/ODI Workshop on commons, wastelands, trees and the poor: finding the right fit, Institute of Development Studies at the University of Sussex, June 1987.

Ninez, V.K. (1984). *Household gardens: theoretical considerations on an old survival strategy*, Potatoes in Food Systems Research Series, Report No. 1. International Potato Center, Lima, Peru.

Raintree, J.B. (1989). *Social, economic and political aspects of agroforestry*. Paper presented at the International Conference on Agroforestry: principles and practice, Edinburgh, July 1989.

Raintree, J.B. (1991). *Socioeconomic attributes of trees and tree planting practices*, Community Forestry Note No. 9. FAO, Rome.

Shepherd, G. (1989). Assessing farmers' tree-use and tree-planting priorities: a report to guide the ODA/Government of Kenya Embu-Meru-Isiolo Forestry Project. Overseas Development Institute, London.

Shepherd, G. (1992). *Managing Africa's tropical dry forests: a review of indigenous methods*, ODI Agricultural Occasional Paper No. 14. Overseas Development Institute, London.

Tamang, D. (1993). *Living in a fragile ecosystem: indigenous soil management in the hills of Nepal*. Gatekeeper Series No. 41. IIED, London.

Taylor, G.F. and Soumaré M. (1983). Strategies for forestry development in the semi-arid tropics: lessons from the Sahel. In *Strategies and designs for aforestation, reforestation and tree planting*, Proceedings of an international symposium (ed. K.F. Wiersum). PUDOC, Wageningen.

Warner, K. (1993). *Patterns of farmer tree growing in eastern Africa: a socio-economic analysis*, Tropical Forestry Paper No. 27. Oxford Forestry Institute and the International Centre for Research in Agroforestry, Oxford.

Index